Soil Contamination

Soil Contamination

Editor

Aleix Campo

scitus
academics

Soil Contamination
Edited by **Aleix Campo**

ISBN: 978-1-68117-536-2
Library of Congress Control Number: 2016942936

SCITUS Academics LLC,
www.scitusacademics.com
Box No. 4766, 616 Corporate Way,
Suite 2, Valley Cottage,
NY 10989

Notice

Reasonable efforts have been made to publish reliable data and views articulated in the chapters are those of the individual contributors, and not necessarily those of the editors or publishers. Editors or publishers are not responsible for the accuracy of the information in the published chapters or consequences of their use. The publisher believes no responsibility for any damage or grievance to the persons or property arising out of the use of any materials, instructions, methods or thoughts in the book. The editors and the publisher have attempted to trace the copyright holders of all material reproduced in this publication and apologize to copyright holders if permission has not been obtained. If any copyright holder has not been acknowledged, please write to us so we may rectify.

Printed in United States of America on Acid Free Paper

Preface

In recent decades, soil contamination has improved dramatically, mainly due to petroleum hydrocarbons, pesticides, solvents and other heavy metals from human activities and industrial waste. Soil contamination or possibly soil pollution is caused by the presence of xenobiotic (human-made) chemicals or possibly other changes in the organic soil. Sustainable efforts are required to develop good risk assessment procedures, sustainable land management and remediation strategies. Contaminated or possibly polluted soil directly affects human health through direct contact with the soil or possibly through inhalation of vaporized soil contaminants; the infiltration of soil contamination into groundwater aquifers used for human consumption, sometimes in areas apparently far removed from any obvious source of above-ground contamination, poses potentially greater threats. It is not by chance that soil contaminants can have significant adverse effects on ecosystems.

Soil Contamination offers a comprehensive overview of contaminated soil surveillance and remediation methods. Soil is an inimitable resource that supports life on earth, challenged by the demands of a growing population for food and energy. Soil contamination created a crucial problem to be solved if we are to ensure the quality of life of future and present generations. The book covers topics related to the monitoring of dioxin, furan, heavy metals and hydrocarbons in soils-biomarkers and bioindicators for soil toxicity assessment-use of reflectance spectroscopy for soil contaminants and detection of waste materials-remediation technologies and methods.

Table of Contents

Chapter 1

Soil Contamination, Risk Assessment and Remediation

Muhammad Aqeel Ashraf[1], Mohd. Jamil Maah[2] and Ismail Yusoff[1]

[1] Department of Geology, Faculty of Science, University of Malaya, Kuala Lumpur, Malaysia

[2] Department of Chemistry, Faculty of Science, University of Malaya, Kuala Lumpur, Malaysia

1. INTRODUCTION

Environment pollution is a burning topic of the day. Air, water and soil are being polluted alike. Soil being a "universal sink" bears the greatest burden of environmental pollution. It is getting polluted in a number of ways. There is urgency in controlling the soil pollution in order to preserve the soil fertility and increase the productivity. Pollution may be defined as an undesirable change in the physical, chemical and biological characteristics of air, water and soil which affect human life, lives of other useful living plants and animals, industrial progress, living conditions and cultural assets. A pollutant is something which adversely interfere with health, comfort, property or environment of the people. Generally most pollutants are introduced in the environment by sewage, waste, accidental discharge or else they are by-products or residues from the production of something useful. Due to this our precious natural resources like air, water and soil are getting polluted.

The basis of agriculture is Soil. All crops for human food and animal feed depend upon it. We are losing this important natural resource by the accelerated erosion 10 some extent. In addition to this the enormous quantities of man-made waste products, sludge and other product" from new waste treatment plants even polluted water are also causing or leading to soil pollution. In order to preserve the fertility and the productivity of the soil, control measures are to be taken in a herculean manner, thereby improving the health of all living beings.

Assessing the ecological risk of contaminated soil, pesticide application, sewage sludge amendment, and other human activities leading to exposure of the terrestrial environment to hazardous substances is a complicated task with numerous associated problems. Not only is terrestrial ecological risk assessment a relatively new field of science that has developed rapidly only since the mid-1980s, but it is also complicated by the fact that soil, in contrast to most aquatic environments, is very often on private lands and traded as real estate. Professional and economic divergence between the interests of scientists, stakeholders, authorities, engineers, managers, lawyers, nongovernment organizations (NGOs) and regulators is therefore not unusual. Even neglecting those aspects, a number of unresolved problems exist in the way we currently assess risk and manage the impact of anthropogenic substances in the terrestrial environment.

This chapter does not intend to present a comprehensive review of all published data from ecological studies at contaminated sites. Instead, the observations from all case studies are used in the discussion and form the basis for the final conclusion. In each case, we try to answer the following questions:

1 What is soil pollution and how it occurs?

2 How to determine the ecological risk assessment of the soil?

3 To what extent do soil screening levels (over)estimate risk?

4 Do bioassays represent a more realistic risk estimate?

5 Is it possible to make sound field surveys, or do we lack suitable reference situations?

6 What are the possible soil management methods for the polluted soils?

2. SOIL POLLUTION

Soil pollution is defined as the build-up in soils of persistent toxic compounds, chemicals, salts, radioactive materials, or disease causing agents, which have adverse effects on plant growth and animal health [1].

Soil is the thin layer of organic and inorganic materials that covers the Earth's rocky surface. The organic portion, which is derived from the decayed remains of plants and animal, is concentrated in the dark uppermost topsoil. The inorganic portion made up of rock fragments, was formed over thousands of years by physical and chemical weathering of bedrock. Productive soils are necessary for agriculture to supply the world with sufficient food [2].

There are many different ways that soil can become polluted, such as:

- Seepage from a landfill
- Discharge of industrial waste into the soil
- Percolation of contaminated water into the soil
- Rupture of underground storage tanks
- Excess application of pesticides, herbicides or fertilizer
- Solid waste seepage

The most common chemicals involved in causing soil pollution are:

- Petroleum hydrocarbons
- Heavy metals
- Pesticides
- Solvents

2.1. Inorganic Toxic Compounds

Inorganic residues in industrial waste cause serious problems as regards their disposal. They contain metals which have high potential for toxicity. Industrial activity also emits large amounts of arsenic fluorides and sulphur dioxide (SO_2)

[3]. Fluorides are found in the atmosphere from superphosphate, phosphoric acid, aluminium, steel and ceramic industries. Sulphur dioxide emitted by factories and thermal plants may make soils very acidic. These metals cause leaf injury and destroy vegetation.

Copper, mercury, cadmium, lead, nickel, arsenic are the elements which can accumulate in the soil, if they get entry either through sewage, industrial waste or mine washings. Some of the fungicides containing copper and mercury also add to soil pollution. Smokes from automobiles contain lead which gets adsorbed by soil particles and is toxic to plants. The toxicity can be minimized by building up soil organic matter, adding lime to soils and keeping the soil alkaline [4].

2.2. Organic Wastes

Organic wastes of various types cause pollution hazards. Domestic garbage, municipal sewage and industrial wastes when left in heaps or improperly disposed seriously affect health of human beings, plants and animals [5-7]. Organic wastes contain borates, phosphates, detergents in large amounts. If untreated they will affect the vegetative growth of plants. The main organic contaminants are phenols and coal.

Asbestos, combustible materials, gases like methane, carbon dioxide, hydrogen sulphide, carbon monoxide, sulphur dioxide, petrol are also contaminants. The radioactive materials like uranium, thorium, strontium etc. also cause dangerous soil pollution. Fallout of strontium mostly remains on the soil and is concentrated in the sediments [8]. Decontamination procedures may include continuous cropping and use of chelate amendments. Other liquids wastes like sewage, sewage sludge, etc. are also important sources of soil problems.

- Sewage and sewage sludge

Soil pollution is often caused by the uncontrolled disposal of sewage and other liquid wastes resulting from domestic uses of water, industrial wastes containing a variety of pollutants, agricultural effluents from animal husbandry and drainage of irrigation water and urban runoff [9-10]. Irrigation with sewage water causes profound changes in the irrigated soils. Amongst various changes that are brought about in the soil as an outlet of sewage irrigation include physical changes like leaching, changes in humus content, and porosity etc., chemical changes like soil reaction, base exchange status, salinity, quantity and availability of nutrients like nitrogen, potash, phosphorus, etc. Sewage sludges pollute the soil by accumulating the metals like lead, nickel, zinc, cadmium, etc. This may lead to the phytoxicity of plants.

- Heavy metal pollutants

Heavy metals are elements having a density greater than five in their elemental form. They mostly find specific absorption sites in the soil where they are retained very strongly either on the inorganic or organic colloids. They are widely distributed in the environment, soils, plants, animals and in their tissues. These are essential for plants and animals in trace amounts. Mainly urban and industrial aerosols, combustion of fuels, liquid and solid from animals and human beings, mining wastes, industrial and agricultural chemicals etc. are contributing heavy metal pollution. Heavy metals are present in all uncontaminated soils as the result of weathering from their parent materials. Concentration of heavy metals in soils and plants is given in Table 1.

Table 1. Heavy metal concentration in the hithosphere, soils and plants (Ug/gm dry matter)

1	Cadmium (Cd)	0.2	0.01-0.7	0.2-0.8
2	Cobalt (Co)	40	1-40	0.05-0.5
3	Chromium (Cr)	200	5-3000	0.2-1.0
4	Copper (Cu)	70	2-100	4-15
5	Iron (Fe)	50,000	7000-5,50,000	140
6	Mercury (Hg)	0.5	0.01-0.3	0.015
7	Manganese (Mn)	1000	100-4000	15-100
8	Molybdenum (Mo)	2.3	0.2-5	1-10
9	Nickel (Ni)	100	10-1000	1
10	Lead (Pb)	16	2-200	0.1-10
11	Tin (Sn)	40	2-100	0.3
12	Zinc (Zn)	80	10-300	8-100

In agricultural soils, however, the concentration of one or more of these elements may be significantly increased in several ways, like through applications of chemicals, sewage sludge, farm slurries, etc. Increased doses of fertilizers, pesticides or agricultural chemicals, over a period, add heavy metals to soils which may contaminate them. Certain phosphatic fertilizers frequently contain trace amounts of cadmium which may accumulate in these soils.

Likewise, some fertilizers when applied to soils, they add certain heavy metals which are given in Table 2.

Table 2. Heavy metal content of fertilizers (ug/gm)

Sl.No	Fertilizer	Co	Cr	Cu	Mn	Mo	Ni	Pb	Zn
1	Nitrochalk	-	-	22	24	-	2	-	15
2	Calcium	0.1	Traces	Traces	Traces	-	-	-	1
3	Nitrate	-	-	To 10	To 5	-	-	-	-
4	Ammonium sulphate	<5	<5	0.800	0.80	<0.05 to 0.22	<5	Traces to 200	0.800
5	Super phosphate	0.02-13	0-1000	Traces to 1000	Traces to 2842	Traces to 35	Traces to 32	Traces to 92	70-3000
6	Potassium chloride	001	-	0-10	Traces-8	<0.05	<1	<1	0-3
7	Potassium sulphate	<5	<5	0-300 to 80	Traces to 33	0.09	<5	<50	<50

The range of heavy metal contents in sludges is given in Table 3.

Table 3. Heavy metal contents in sludges (ppm)

Sl.No	Heavy metal	Range (ppm)
1	Cadmium	< 60-1500
2	Cobalt	2-260
3	Chromium	40-8800
4	Copper	200-8000
5	Iron	6000-62,000
6	Manganese	150-2500
7	Molybdenum	2-30
8	Nickel	20-5300
9	Lead	120-3000
10	Zinc	700-49,000

The fate of heavy metals in soil will be controlled by physical and biological processes acting within the soil. Metal ions enter the soil solution from these various forms of combination in different rates they may either remain in solution or pass into the drainage water or be taken up by plants growing on the soil or be retained by the soil in sparingly soluble or insoluble forms. The organic matter of these soil have great affinity to heavy metals cations which form stable complexes thereby leading to reduced nutrient content [11-12].

2.3. Organic Pesticides

Pesticides are quite frequently used to -control several types of pests now-a-days. Pesticides may exert harmful effects to micro-organisms, as a result of which plant growth may be affected. Pesticides which are not rapidly decomposed may create such problems. Accumulation is residues of pesticides in higher concentrations are toxic. Pesticides persistence in soil and movement into water streams may also lead to their entry into foods and create health hazards. Pesticides particularly aromatic organic compounds are not degraded rapidly and therefore, have a long persistence time which can be seen inTable 4.

Table 4. Persistence time for some selected pesticides

Sl.No	Pesticide	Persistence time
1	BHC	11 yrs
2	DDT	10 yrs
3	2,4-D	2-8 weeks
4	Aldrin	9 yrs
5	Diuron	16 months
6	Atrazine	18 months
7	Siwazine	17 months
8	Chlordane	12 yrs
9	2,3 6-Trichlorobenzene (TBA)	2-5 yrs

Mercury, cadmium and arsenic are common constituents of pesticides and all these heavy metals are toxic. At present DDT and a number of organochlorine compounds used as pesticides have been declared harmful and banned in U.S.A. and England [13-14]. It is due to the persistence of their residues in soils for considerable time without losing their toxicity. This has led to higher concentration of these pesticides in vegetation, in animal flesh and milk. Eventually man has been affected. In view of their demerits, organochlorines have been replaced by organophosphate pesticides which are more toxic, but do not leave any residue. They do not pollute the soil. The rodenticides too add to soil pollution. A major method of checking this pesticidal pollution is to increase the organic matter content of the sol and choose such pesticides which are non-persistent and leave no harmful residue.

2.4. Types Of Soil Pollution

- Agricultural Soil Pollution
    - pollution of surface soil
    - pollution of underground soil
- Soil pollution by industrial effluents and solid wastes
    - pollution of surface soil
    - disturbances in soil profile
- Pollution due to urban activities
    - pollution of surface soil
    - pollution of underground soil

2.5. Sources Of Soil Pollution

The sources which pollute the soil are twofold: Agricultural sources and non-agricultural sources.Figure 1 shows the different sources for the soil pollution.

- Agricultural sources

Soil pollution comes from different sources including agriculture and animal husbandry. Some of the agricultural practices lead to soil pollution. They are animal wastes, use of long lived pesticides, herbicides, fungicides, nematocides, etc. fertilizers and some agricultural practices.

- Non-agricultural sources

Soil pollution by non-agricultural sources is usually the direct result of urban sprawl caused by rapidly increasing population and a rapidly per capita output of waste related to our modem way of life. Its materials that find their entry into the soil system have long persistence and accumulate in toxic concentration and thus become sources of pollution. Some of those most important soil pollutants are inorganic toxic compounds.

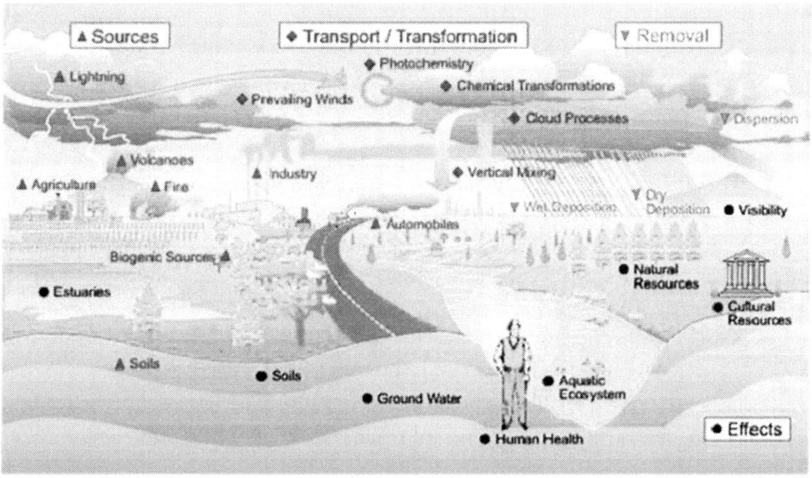

Figure 1. Sources of Soil Pollution [5]

2.6. Causes Of Soil Pollution

Soil pollution is caused by the presence of man-made chemicals or other alteration in the natural soil environment. This type of contamination typically arises from the rupture of underground storage links, application of pesticides, and percolation of contaminated surface water to subsurface strata, oil and fuel dumping, leaching of wastes from landfills or direct discharge of industrial wastes to the soil. The most common chemicals involved are petroleum hydrocarbons, solvents, pesticides, lead and other heavy metals. This occurrence of this phenomenon is correlated with the degree of industrialization and intensities of chemical usage. A soil pollutant is any factor which deteriorates the quality, texture and mineral content of the soil or which disturbs the biological balance of the organisms in the soil. Pollution in soil has adverse effect on plant growth.

Pollution in soil is associated with

- Indiscriminate use of fertilizers
- Indiscriminate use of pesticides, insecticides and herbicides
- Dumping of large quantities of solid waste
- Deforestation and soil erosion

- Indiscriminate use of fertilizers

Oxygen from air and water but other necessary nutrients like nitrogen, phosphorus, potassium, calcium, magnesium, sulfur and more must be obtained from the soil. Farmers generally use fertilizers to correct soil deficiencies. Fertilizers contaminate the soil with impurities, which come from the raw materials used for their manufacture. Mixed fertilizers often contain ammonium nitrate (NH_4NO_3), phosphorus as P_2O_5, and potassium as K_2O. For instance, As, Pb and Cd present in traces in rock phosphate mineral get transferred to super phosphate fertilizer. Since the metals are not degradable, their accumulation in the soil above their toxic levels due to excessive use of phosphate fertilizers becomes an indestructible poison for crops. The over use of NPK fertilizers reduce quantity of vegetables and crops grown on soil over the years. It also reduces the protein content of wheat, maize, grams, etc., grown on that soil. The carbohydrate quality of such crops also gets degraded [15]. Excess potassium content in soil decreases Vitamin C and carotene content in vegetables and fruits. The vegetables and fruits grown on over fertilized soil are more prone to attacks by insects and disease.

- Indiscriminate use of pesticides, insecticides and herbicides

Plants on which we depend for food are under attack from insects, fungi, bacteria, viruses, rodents and other animals, and must compete with weeds for nutrients. To kill unwanted populations living in or on their crops, farmers use pesticides. The first widespread insecticide use began at the end of World War II and included DDT (dichlorodiphenyltrichloroethane) and gammaxene. Insects soon became resistant to DDT and as the chemical did not decompose readily, it persisted in the environment. Since it was soluble in fat rather than water, it biomagnified up the food chain and disrupted calcium metabolism in birds, causing eggshells to be thin and fragile. As a result, large birds of prey such as the brown pelican, ospreys, falcons and eagles became endangered. DDT has been now been banned in most western countries. Ironically many of them including USA, still produce DDT for export to other developing nations whose needs outweigh the problems caused by it [16].

- Dumping of solid wastes

In general, solid waste includes garbage, domestic refuse and discarded solid materials such as those from commercial, industrial and agricultural operations. They contain increasing amounts of paper, cardboards, plastics, glass, old construction material, packaging material and toxic or otherwise hazardous substances. Since a significant amount of urban solid waste tends to be paper and food waste, the majority is recyclable or biodegradable in landfills. Similarly, most agricultural waste is recycled and mining waste is left on site. The portion of solid waste that is hazardous such as oils, battery metals, heavy metals from smelting industries and organic solvents are the ones we have to pay particular attention to. These can in the long run, get deposited to the soils of the surrounding area and pollute them by altering their chemical and biological properties [17].

- Deforestation

Soil Erosion occurs when the weathered soil particles are dislodged and carried away by wind or water. Deforestation, agricultural development, temperature extremes, precipitation including acid rain, and human activities contribute to this erosion. Humans speed up this process by construction, mining, cutting of timber, over cropping and overgrazing. It results in floods and cause soil erosion. Forests and grasslands are an excellent binding material that keeps the soil intact and healthy. They support many habitats and ecosystems, which provide innumerable feeding pathways or food chains to all species. Their loss would threaten food chains and the survival of many species. During the past few years quite a lot of vast green land has been converted into deserts. The precious rain forest habitats of South America, tropical Asia and Africa are coming under pressure of population growth and development (especially timber, construction and agriculture). Many scientists believe that a wealth of medicinal substances including a cure for cancer and aids, lie in these forests. Deforestation is slowly destroying the most productive flora and fauna areas in the world, which also form vast tracts of a very valuable sink for CO_2 [18].

- Pollution due to urbanization

Pollution of surface soils materials (like vegetables, animal wastes, papers, wooden pieces, carcasses, plant twigs, leaves, cloth wastes as well as sweepings) and many non-biodegradable materials (such as plastic bags, plastic bottles, plastic wastes, glass bottles, glass pieces, stone / cement pieces) [19-20]. On a rough estimate Indian cities are producing solid city wastes to the tune of 50,000 - 80,000 metric tons every day. If left uncollected and decomposed, they are a cause of several problems such as;

- Clogging of drains: Causing serious drainage problems including the burst / leakage of drainage lines leading to health problems.
- Barrier to movement of water: Solid wastes have seriously damaged the normal movement of water thus creating problem of inundation, damage to foundation of buildings as well as public health hazards.
- Foul smell: Generated by dumping the wastes at a place.
- Increased microbial activities: Microbial decomposition of organic wastes generate large quantities of methane besides many chemicals to pollute the soil and water flowing on its surface
- When such solid wastes are hospital wastes they create many health problems: As they may have dangerous pathogen within them besides dangerous medicines, injections.

- Pollution of underground soil

Underground soil in cities is likely to be polluted by

- Chemicals released by industrial wastes and industrial wastes
- Decomposed and partially decomposed materials of sanitary wastes

Many dangerous chemicals like cadmium, chromium, lead, arsenic, selenium products are likely to be deposited in underground soil. Similarly underground soils polluted by sanitary wastes generate many harmful chemicals. These can damage the normal activities and ecological balance in the underground soil

2.7. Effects Of Soil Pollution

- Agricultural
  - Reduced soil fertility
  - Reduced nitrogen fixation

  - Increased erodibility
  - Larger loss of soil and nutrients
  - Deposition of silt in tanks and reservoirs
  - Reduced crop yield
  - Imbalance in soil fauna and flora
- Industrial
  - Dangerous chemicals entering underground water
  - Ecological imbalance
  - Release of pollutant gases
  - Release of radioactive rays causing health problems
  - Increased salinity
  - Reduced vegetation
- Urban
  - Clogging of drains
  - Inundation of areas
  - Public health problems
  - Pollution of drinking water sources
  - Foul smell and release of gases
  - Waste management problems
- Environmental
  - Soil becomes unavailable to grow food
  - If contaminated soil is used to grow food, the land will usually produce lower yields

- Can cause even more harm because a lack of plants on the soil will cause more erosion

- The pollutants will change the makeup of the soil and the types of microorganisms that will live in it.

- Thus it's possible for soil pollution to change whole ecosystems

2.8. Control Of Soil Pollution

The following steps have been suggested to control soil pollution. To help prevent soil erosion, we can limit construction in sensitive area. In general we would need less fertilizer and fewer pesticides if we could all adopt the three R's: Reduce, Reuse, and Recycle. This would give us less solid waste.

- Extraction and separation techniques

In solvent extraction, the contaminated & oil is mixed with an extracting agent in general (an aqueous solution but preferably an organic solvent). Potential applications include the removal of metals such as cadmium, copper, zinc, nickel, chromium, arsenic, antimony and lead using a mineral solution, zinc lead, organo-metallic compounds and some cyanides using sodium hydroxide solution. Hydrocarbons and halogenated hydrocarbons can also be removed [21]. Contamination is often preferentially present in the finer or coarser fraction of the soil or the organic components (ex. humus), contamination can therefore, be removed in some cases using a process which separates the soil into fractions on the basis of specific gravity or particle size or settling velocity.

- Thermal methods

In thermal methods, there are two ways of heat treatment; removal of contaminants by evaporation either by direct heat transfer from heated air or an open flame or by indirect heat transfer, and destruction of the contaminants directly or indirectly at an appropriate temperature. The gas leaving the heating appliance must be treated to destroy or remove any contaminants or unwanted products of combustion. A related process is stream stripping in which steam is injected into soil to aid evaporation of relatively volatile contaminants which may be water soluble or insoluble.

- Chemical methods

Treatment of the soil in suspension in a suitable liquid and without sludging is the two possible methods. In these, intimate, contact between soil and chemical is essential and should be frequently done so that the process of detoxification is complete.

- Microbial treatment methods

The microbial treatment methods appear to be more promising which can deal with whole range of organic contaminants including phenol, polychlorinated hydrocarbons, oil and oil products, dioxins, etc. There are two different ways of approaching the problems.

1 A community of microbes already existing on the site is collected and cultured in the laboratory.

2 Strains of microbes are developed in the laboratory that is capable of metabolizing particular chemicals.

Excavation of the soil prior to treatment offers the greatest scope for creating optimum conditions. The excavated soil can be placed on thin layers to various depth using standard earth moving techniques and microbes and nutrients applied using standard agricultural techniques such as fertilizing, ploughing, harrowing, etc.

Reducing chemical fertilizer and pesticide use

Applying bio-fertilizers and manures can reduce chemical fertilizer and pesticide use. Biological methods of pest control can also reduce the use of pesticides and thereby minimize soil pollution.

Reusing of materials

Materials such as glass containers, plastic bags, paper, cloth etc. can be reused at domestic levels rather than being disposed, reducing solid waste pollution.

Recycling and recovery of materials

This is a reasonable solution for reducing soil pollution. Materials such as paper, some kinds of plastics and glass can and are being recycled. This decreases the volume of refuse and helps in the conservation of natural resources. For example, recovery of one tonne of paper can save 17 trees.

Reforesting

Control of land loss and soil erosion can be attempted through restoring forest and grass cover to check wastelands, soil erosion and floods. Crop rotation or mixed cropping can improve the fertility of the land.

Solid waste treatment

Proper methods should be adopted for management of solid waste disposal. Industrial wastes can be treated physically, chemically and biologically until they are less hazardous. Acidic and alkaline wastes should be first neutralized; the insoluble material if biodegradable should be allowed to degrade under controlled conditions before being disposed. As a last resort, new areas for storage of hazardous waste should be investigated such as deep well injection and more secure landfills. Burying the waste in locations situated away from residential areas is the simplest and most widely used technique of solid waste management.

2.9. Soil Monitoring

The main objective of soil monitoring is to prevent and mitigate contamination by substances with the potential to exert an adverse effect on the soil itself, and on air, water and organisms that may contact the soil. Soil monitoring, within the approvals program, is directed primarily to the assessment of contaminants that have been released to the soil surface. Thus, subsurface facilities are generally not the reason for soil monitoring, but may be the reason for groundwater monitoring [22]. However, where soil contamination is known or suspected to originate from subsurface sources such as underground tanks or pipes, an assessment will be required. Where the above considerations indicate soil monitoring is required as a condition of an Approval, the proponent is required to carry out the following, as specified in the Soil Monitoring Directive: > prepare a soil monitoring proposal; > execute the approved soil monitoring plan; > interpret and report the results of the soil monitoring; and > prepare and execute a soil management plan where indicated by the results of soil monitoring. This guideline provides a background for the soil monitoring program and a description of soil management program requirements.

Legislative background

The soil monitoring program mostly developed under the Environmental Protection in support of the following principles: > development must be sustainable, meaning that the use of resources and the environment today must not impair prospects for their use by future generations; > the environmental impact of development must be prevented or mitigated; > polluters should bear the responsibility of paying for the costs of their actions; > remediation costs should be incorporated into financial planning so that adequate funds are available for site remediation and planners can know the true costs and benefits of source reduction programs. Recognizing that under the environmental protection is a shared responsibility, it follows that both the approval holder and the Department must have a means to assess environmental performance with respect to the above principles and requirements.

Soil quality standards

Environmental Protection expects that approval holders will manage their operations to prevent substance releases to soil. Substance releases to soil do occur, however, and contaminants are often present above background concentrations at industrial facilities. In view of this, Environmental Protection should have soil quality standards to guide assessment and remediation of soil contamination. Facilities that are currently uncontaminated have the opportunity to maintain conditions that allow unrestricted land-use. For these facilities, the minimum standards will be determined by the Tier I criteria or equivalent objectives. Older facilities, however, were often operated under different standards and environmental management practices than are currently acceptable.

2.10. Soil Pollution Risks

There is an increasing use of risk-oriented policies to deal with the local effects of soil pollution. The risks that such policies deal with are: human health risks and can also include eco-toxicological risks. These risks are expressed in terms of negative effects and chances between 0 and 1 that such negative effects will occur. Examples of areas where risk-oriented policies are applied to soil pollution include the United States of America [23], Canada [24] and countries in the European Union [25]. Historically, these risk oriented policies have followed the abandonment of policies aimed at restoring soils to their original 'clean' state.

Risk-based criteria or standards, developed in the framework of risk oriented policies, are applied to risks estimated with deterministic methodologies, following the steps of hazard characterization, appraisal of exposure and risk characterization, while using exposure-risk relations established beforehand. Risk-based criteria have been applied to decisions about soil remediation in the form of soil clean-up standards [26], to the use of soils for specific purposes and in the United States also to sediment management [27]. The risk-oriented policies considered here [28], assume that background exposure to pollutants carries no risk and that a specified level of soil pollution carries a maximum tolerable or maximum acceptable risk for organisms living locally. The latter is the main basis for standard setting.

In part, risk-oriented soil pollution legislation includes policy goals that are qualitative [29]. For instance, the primary UK legislation on contaminated soil defines land as contaminated in need of risk management 'if significant harm is being caused or there is a significant possibility of such harm being caused' [29]. Mostly, however policies have resulted in specific quantitative values for maximum tolerable or acceptable soil pollution. The analysis of such values used in different industrialized countries has shown that there are very large differences, roughly up to a factor [30]. According to Provoost et al. [31], these differences to a large extent originate in different political choices (e.g. including or excluding ecotoxicity) and in different assumptions as to the modeling of exposure to soil pollutants, including site related factors, such as soil type and building constructions [31].

2.10.1. Risks Related To One Soil Pollutant

In practice, there are several matters which are at variance with the proper establishment of actual risk related to one soil pollutant. These are: the absence of standards for pollutants, neglect of background exposure, and neglect of routes of exposure to soil pollution, neglect of available dose- effect studies and neglect of biological availability. These will now be discussed in more detail.

Absence of quality standards

When data regarding soil pollutants are available, they should be compared with quality standards reflecting maximum tolerable risk of exposure. However, such standards are not always in place. For instance, of the volatile organic carbon compounds detected in groundwater samples by the US Geological Service, were unregulated- with no standards in place [32]. Similarly Patterson et al. [33] found a variety of brominated ethenes in Australian groundwater, all lacking standards.

Neglect of background exposure

For a proper estimate of soil pollution related risks, exposure to specific soil pollutants should be evaluated in combination with exposure to the same substance that is not related to local soil contamination. Several countries, such as Canada, Germany, Spain and Belgium, do indeed establish soil clean-up standards while considering background dietary and inhalatory exposure but others, e.g. Sweden, Norway and the Netherlands, do not [34]. Neglecting background exposure or specific types of background exposure may have implications for risk estimates.

Neglect of routes of exposure to soil pollution

In evaluating exposure to soil pollutants, assumptions regarding exposure routes are important. In this respect difference between countries may be noted. Soil clean-up standards for lead of Norway and Sweden differ in part because in Sweden the dominant exposure route is assumed to be by drinking water and in Norway it is thought to be by drinking water and ingestion of soil [34].

Inhalation of household dust and soil particles is not always taken into account in governmental decision making about risks of soil pollution. For instance, in the Netherlands inhalation of soil particles has been neglected as an exposure route, but in e.g. Spain it is not [34]. Neglect of inhalation would seem at variance with existing studies. Nawrot et al. [35] have studied the effects of cadmium pollution in soil (around former thermal zinc plants) and found a significant increase in lung cancer risk correlated with cadmium exposure. They plausibly explain this in terms of exposure of lung tissue to cadmium present in inhaled soil and household dust particles.

Household dust particles have also been found to be important in the exposure of children to pesticides in agricultural settings.

Neglect of available dose-effect studies

Akesson et al. [37] have analyzed the effects of low environmental cadmium exposure in an epidemiological study of Swedish women in the Lund area, being 53-64 years of age, excluding women from areas with soils heavily polluted by cadmium. Akesson et al. [37] found associations between the internal dose of cadmium and tubular and glomerular kidney effects, which may represent early signs of adverse effects. Women with diabetes seemed to be at increased risk of experiencing such early signs. In view of these data it seems plausible that at a background exposure that is common in Sweden, old women in the general population may be at risk for adverse cadmium effects [34] and that even a modest increase in cadmium exposure due to polluted soil may lead to added risk. However, when establishing soil clean-up standards in Sweden this background exposure has been neglected [34]. Nawrot et al. [35] have studied the relation between mortality and cadmium body burden in Belgium. They obtained evidence that total mortality and non-cardiovascular mortality may be elevated at cadmium body burdens which can be found among the population not living on soils that are currently considered to be a health risk. Similarly there are now strong indications that the negative effects of lead on the neurophysiologic and sexual development may well be found at the level of background exposure common in Western European and US cities [37], though soil pollution policy, at least in European counties, assumes that such background exposure is safe [14].

2.10.2. Ecotoxicological Risks

Maximum acceptable or maximum tolerable ecotoxicological risks are usually derived from a limited number of studies concerning single species under laboratory conditions. Laboratory conditions may be very different from actual conditions in the field, and thus findings in the field are often at variance with laboratory studies [35]. In field studies it has been found that several factors which tend to be neglected in laboratory studies may strongly impact toxic effects of soil pollutants. These include among others: density and adaptability of populations of affected organisms, the presence of other environmental stress factors and the presence or absence of specific landscape elements such as buffer strips [35].

2.10.3. Biological Availability

Biologically available pollutants determine risk [3]. Biological availability may vary strongly for different types of organisms [36]. Biological availability of a compound in a specific soil is also dependent on physical, chemical and biological and spatial factors [35]. Examples of such factors are pH, the amount and nature of organic and mineral compounds also present and the presence of organisms that can mobilize soil pollutants [37-39]. In practice, biological availability may be much at variance with total concentrations [40].

2.10.4. Combination Effects

Limited accounting of combination effects

As to the overall risk of soil pollutants, cumulative effects of the combination of substances present in soils should be considered. However actual standard setting practice has largely focused on criteria relating to one element or compound. In some cases there are criteria for groups of compounds [34]. Such criteria limit the amount (in g/kg soil) of groups of compounds but often do not address the possibility that the risk per unit of weight may be different for different compounds. An exception to this is criteria for the presence of halogenated dioxins and benzofurans and planar biphenyls. The establishment of risk in case of exposure to these compounds uses addition on the basis of equivalent toxicity [40]. This is a major improvement, though it has been pointed out that this approach may still underestimate the risk of neurodevelopment effects [41].

Importance of combination effects

Combination effects may be important in two respects. Firstly, coexisting soil contaminants may impact each others' biological availability [43]. Secondly, exposure to a combination of pollutants may be associated with antagonistic, synergistic and additive interactions of these pollutants, impacting their effect on organisms [44-47]. Some risks of pollutant mixtures can be predicted on the basis of existing knowledge. For instance there is a fair chance that there will be dose additivity when effects are receptor mediated [48]. Also in case of narcotic effects, joint-mixture ecotoxicological effects may be predicted [48]. If responses are dissimilar, response addition may be used [49]. A methodology to deal with the ecotoxicity of mixtures giving rise to both dose-additive and response-additive effects has been proposed [48]. This two step model evaluates mixture toxicity for the same mode of action with concentration additivity and the toxicity for different modes of action with response additivity. For determining the severity of ecotoxicological effects in case of heavily polluted soils (in which legal maximum tolerable levels for one or more substances are exceeded), a systematic approach to combination effects based on a mixture of concentration addition and response addition has been proposed [41].

2.10.5. Remedies For Shortcomings

Remedies would seem possible which would allow for a significant improvement in risk estimates. Unregulated substances can get standards. Standards may be regularly updated on the basis of new dose-effect studies. Risk estimates can include both background exposure and all exposure routes for local soil pollution. Estimates of biological availability can be integrated in risk assessments and improved by better testing of bioavailability or by in-vivo monitoring [51]. The deficiencies in taking account of combination effects in ecotoxicity, discussed in section may be addressed by directly testing of ecotoxicity, when the focus is on ecosystem functioning [52,53]. However it should be noted that small effects on the functioning of ecosystems may have large effects over time [35]. This necessitates large numbers of replicate tests that may well be beyond routine practice [35].

In determining combination effects on human health, direct testing on humans is an 'unethical option'. However biomarker-based monitoring of some aspects of soil pollution relevant to humans may be an option. For instance Roos et al. [54] have applied a biomarker based test to original and remediated soils that were contaminated by a variety of polycyclic aromatic hydrocarbons (PAH). They tested the expression profile of cytochromes P 450 [54]. Xiao et al. [55] have measured genotoxic risk of soil contamination using an in-vitro assay with Salmonella. Though the relation between such biomarker-based data gathered and the in-vivo risks awaits further elucidation, the application of tests based on biomarkers for soil pollution is an interesting option in dealing with combination effects on humans.

Also, estimates of risk may be derived from biomarkers which may be monitored in people exposed to soil pollution. Such biomarkers have emerged from epidemiological studies considering the combined effect of substances. An illustration thereof is the study by Lee et al. [56] which found a graded association of the concentration of blood lead and urinary cadmium concentrations with oxidative stress related markers in the US population. This suggests that oxidative stress may be useful as a biomarker for combination effects. It has furthermore been proposed to evaluate effects of exposure to nitroarenes by measuring haemoglobin adducts [57], and of mixtures of volatile organochlorines by measuring glutathione conjugative metabolites [58]. Bioassays based on aryl hydrocarbon (Ah) receptor mediated mechanisms have been proposed which will allow a better alternative to the measurement of polyhalogenated aromatic hydrocarbons [41]. Another option is to estimate risks to human health by taking into account cumulative combination effects in line with established cause-effect relations and research into the effects of actual combinations. It has been shown that risks of compounds with the same targets and the same modes of action may be estimated on the basis of concentration addition, while including toxicity equivalence factors for the compounds involved [59].

This has been shown to apply to receptor-mediated-and reactive mechanisms of toxicity, provided that no chemical reactions occur between the components of the mixture considered [60]. Currently this approach is applied to halogenated dioxins, benzofurans and planar polybiphenyls, though non-linear interactions are not completely absent in this category of compounds [61], and neurodevelopment effects may be underestimated, as pointed out before [41]. Extension of this approach is possible to e.g. polycyclic aromatics, including heterocyclic polycyclic aromatics [18,62] organophosphates that inhibit the enzyme cholinesterase [44,63], compounds that bind to estrogen receptors [64-66], carcinogens [67], a variety of petroleum products [68] and compounds that inhibit the MXR efflux pump [69].

3. ECOLOGICAL RISK ASSESSMENT

Ecological risk assessment (ERA) is a process of collecting, organizing, and analyzing environmental data to estimate the risk or probability of undesired effects on organisms, populations, or ecosystems caused by various stressors associated with human activities. The basic principles of ecological risk assessment are described in numerous papers [70-72]. All varieties of ERA are associated with uncertainties. The value or usefulness of the different ERA methodologies depends on the uncertainty, predictability, utility, and costs. There are typically two major types of ERA. The first is predictive and is often associated with the authorization and handling of hazardous substances such as pesticides or new and existing chemicals in the European Union. This kind of ERA is ideally done before environmental release. The second type of ERA could be described as an impact assessment rather than a risk assessment, as it is the assessment of changes in populations or ecosystems in sites or areas already polluted. The predictive method is based on more or less generic extrapolations from laboratory or controlled and manipulated semi field studies to real-world situations. The descriptive method is more site specific as it tries to monitor ecosystem changes in historically contaminated soils such as old dumpsites or gas facilities or in field plots after amendment with pesticides or sewage sludge, for example.

Often ERA is performed in phases or tiers, which may include predictive as well as descriptive methods. The successive tiers require, as a rule of thumb, more time, effort, and money. The paradigm or schemes for ERA may vary considerable from country to country, but often consist of an initial problem formulation based on a preliminary site characterization, and a screening assessment, a characterization of exposure, a characterization of effects, and a risk characterization followed by risk management. Although exposure assessment is often just as or even more important, this chapter primarily considers effect assessment. In most European countries, ERA of contaminated soils consists of rather simplified approaches including soil screening levels (SSL) (a.k.a. quality objectives, quality criteria, benchmarks, guideline values) and simple bioassays for a first screening of risk [73-75]. National research or remediation programs have led to the development of a large variety of guideline values.

Although hard to categorize, most fall into two categories: generic or site specific. While the site-specific guidelines require a characterization of pH, organic matter, etc., at the site, generic guideline values are more independent of modifying factors and hence straightforward to legislate. Three major classes of tools for assessing ecological effects may be identified: standardized ecotoxicity experiments with single species exposed under controlled conditions to single chemicals spiked to soil; *ex situ*bioassays, here defined as simple laboratory assays where single species are exposed to historically contaminated soils collected in the field; and finally monitoring, analyzing, and mapping of population or community structures in the field. Furthermore, mesocosm, lysometer, or terrestrial model ecosystems (TME) may be useful; these may be considered as large (multispecies) bioassays or ecotoxicity tests [76-79]. TMEs have the advantage that they operate with the (relatively) undisturbed intrinsic soil populations that make up a small food web. TME hence allow the assessment of effects of toxicants that are mediated through changes in food supply or competition and predation.

One of the keystones in deriving environmental quality criteria is the use of standardized terrestrial test procedures. The emphasis of these prognostic tests is on reproducibility, standardization, international acceptance, and site independence. Although increasing in numbers, relatively few terrestrial tests are still approved by the International Standardisation Organisation (ISO) or Organization for Economic Cooperation & Development (OECD). However, other tests have shown promising results and are likely to be prepared for standardization in the future [80].

However, the major problem in using simple laboratory tests to extrapolate to contaminated land may not be the limitations of test species and the natural variation in species sensitivity. The problems associated with extrapolating from one or a few species, exposed under controlled and typically optimal conditions, to the complex interaction of species and chemicals found in most contaminated ecosystems should also cause concern. Although single-species laboratory tests with spiked materials have their obvious benefits, e.g., they measure direct toxicity of chemicals and interpretation is therefore simple, supplementary tools are often needed. Bioassays, as defined in this context are one of the more frequently used higher-tier alternatives. Basically the same test species may be used in bioassays for assessing the risk of a specific contaminated soil as in standard laboratory tests. However, bioassays have the advantage, compared to the use of spiked soil samples, that the exact toxicity of a specific soil may be accessed directly: this includes the combined and site-specific toxicological effect of the mixture of contaminants and their metabolites. Furthermore, the *in situ* bioavailability of that specific soil is (at least almost) maintained in the laboratory during the exposure period. Several studies have shown a reduction in bioavailability and/or toxicity of soils with an old history of contamination [80-85].

Bioassays are therefore often considered a more realistic tool than generic soil screening levels based on spiked laboratory soils. However, a number of uncertainties or problems may be associated with the use of bioassays and the interpretation of their results. First, the test species are still exposed to the contaminants in a relatively short period compared to the permanent exposure condition found at contaminated sites. Furthermore, they are exposed under more or less optimal conditions, in that stressors such as predation inter- and interspecies competition, drought, frost, and food depletion are eliminated during exposure. Finally, typically only a few species are tested individually.

To compensate for some of the limitations just described, contaminated soil may be assessed using multispecies mesocosms, lysometers, or TME. In these, species interactions may be evaluated by manually introducing several species to the systems or monitoring the intrinsic populations of the soil. Natural climatic conditions may be included if the test system is kept outdoors. However, if we want to get a more realistic and large-scale picture of the impact caused by, for example, pesticide use or sewage sludge application, or to assess the environmental health at waste sites, industrial areas, or gas works, it is often necessary to conduct some kind of field observations. Several case studies exist in which field studies have successfully elucidated the ecological risk of specific activities or the ecological impact at specific sites [85-87]. The small single-species bioassay, large multispecies TME, and field surveys have some drawbacks in common. First of all, it may be difficult to actually link the observed effect to a specific toxic component in the soil. Which of the many substances is actually causing the majority of the observed effects, or is it perhaps a combination of effects? For a hazard classification of soils or a ranking of soils this may not be so important.

However, to evaluate potential risk-reduction measures or risk management procedures it may be important to identify the most problematic substances. A comparison of soil screening values with measured concentrations for each chemical present at a site may be helpful to identify the most likely group of substances causing the observed effect. Other possible tools may include a toxicity identification evaluation (TIE) approach [88]. The TIE approach is a relatively new method, which aims to identify groups of toxicants in soils with mixed pollution. Potentially toxic components present in the soil are fractionated and determined, and the toxicity of each individual fraction is determined by a *Lux* bacteria-based bioassay or the Microtox bioassay. Although perhaps promising, TIE is a time-consuming and hence costly procedure not yet used routinely. Another crucial issue when analyzing the result of bioassays, TME, and field studies is the presence or absence of a proper reference site or soil. The control soil should in principle resemble the contaminated soil in all relevant parameters, e.g., texture, pH, organic matter, waterholding capacity, and nutrient content, a practical problem that very often is difficult to solve. The lack of adequate control or reference sites may, however, be conquered at least partially by the use of multivariate techniques [89], which relate the species composition and abundance to gradients of pollutants. It is not the intention of this chapter to present a review of statistical tools for ecological risk assessment, and hence a detailed discussion about the use of these is not given.

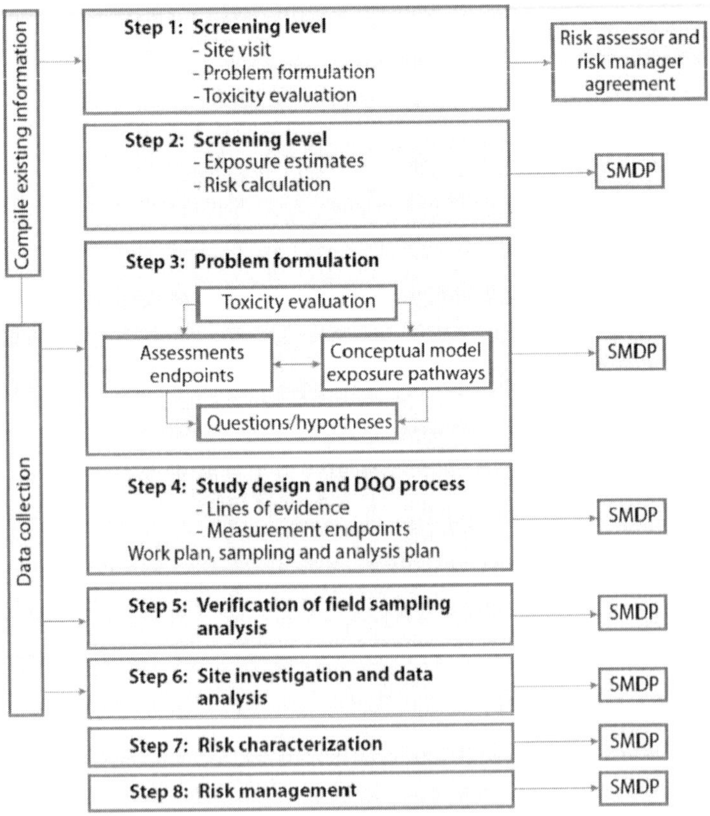

Figure 2. The eight steps in the US-EPA framework for risk assessment of contaminated Superfund sites. DQO = data quality objectives [89]

However, it is obvious that increased computer power and the presence of new easy-to-use software tools have increased the possibility to move away from more conventional univariate statistics such as analysis of variance (ANOVA) to more powerful multivariate statistics that use all collected data to evaluate effects at a higher level of organization. Statistical methods such as the power analysis may also be very useful in planning and designing large-scale ecotoxicity studies such as mesocosms,TME, or field surveys.

The US-EPA has published an Ecological Risk Assessment Guidance, which should be followed when assessing risks at Superfund sites. As all sites are considered unique this should always be done in a site-specific manner. The ERA process suggested by the US-EPA for Superfund sites follows an eight step process, which can be broken down into four categories, i.e. 1) planning and scoping, 2) problem formulation, 3) stressor response and exposure analysis and 4) risk characterisation. Essential for all steps are a negotiation and agreement of the need for further action between the risk assessor, the risk manager and other stakeholders, the so-called scientific-management decision points (SMDP).

SMDP made at the end of the screening-level assessment will not set an initial cleanup goal. Instead, hazard quotients, derived in this step, are used to help determine potential risk. Thus, requiring a cleanup based solely on those values would not be very likely, although it is technically feasible. There are three possible decisions at the SMDP:

1 There is enough information to conclude that ecological risks are very low or non-existent, and therefore there is no need to clean up the site on the basis of ecological risk.

2 The information is not adequate to make a decision at this point, and the ecological risk assessment process will proceed.

3 The information indicates a potential for adverse ecological effects, and a more thorough study is necessary.

In the Netherlands contaminated sites are first determined using a set of soil screening levels called target and intervention values, which take both human and ecological risks into account. At seriously contaminated sites remediation or other soil management decisions are required if the risks cannot be neglected based on a site-specific ecological and human risk assessment, and the chance for dispersion of the contaminants. Until now, the ecological risk assessment has been based on chemical analysis, including a Decision Table harbouring critical dimensions of the impacted area. The United Kingdom and Canada have also developed framework for ecological risk assessment of contamination land. A cornerstone in the UK framework of ERA is the connection to the statutory regime for identification and control of land potentially affected by contamination. The UK framework is based on schemes found in e.g. USA, Canada and the Netherlands. Like these it is a based on a tiered approach where the initial Tier 0 aims to determine whether a site falls under the Part IIA of the legislation. It involves the development of a Conceptual Site Model (CSM), which described what is already (historically) known about the site, e.g. whether there is a likely source-pathway-receptor linkage. The conceptual site model is followed by an initial screening phase (Tier 1) and an actual site-specific characterisation (Tier 2). Tier 1 is a simple deterministic comparison of chemical residue data and the soil quality guideline values supplemented with simple soil-specific toxicity testing. The final step (Tier 3) involves more detailed *in-situ* studies and for example ecological modelling based on a more advanced ecological theory. Tier 3 is not likely to be conducted at many sites.

3.1. Descision Support System For Ecological Risk Assessment

Ecological Risk Assessment is often a complex process with many variables to take into account. ERA involves many stakeholders and all have to be dealt with in a clear and consistent way. A stepwise or tiered approach is therefore useful to overcome the complexity of an ERA. In order to structure all the information collected, a Decision Support System (DSS) can be used. Each tier will lead to a decision to proceed or to stop. A number of decisions supporting systems or frameworks have already been developed in other countries, e.g. UK, the Netherlands and the USA. The DSS presented here is based on basic principles also common in the methodologies used in the USA and UK. However, in the present DSS measures of bioavailability and the use of the Triad approach may be built into the system more systematically. This chapter introduces the overall framework of a novel DSS including the Triad approach and the challenge to weight and scale results used in that process.

3.2. Framework For Ecological Risk Assessment

Rutgers et al. (2000) developed a basic flowchart for Ecological Risk Assessment [90], which is used as the backbone of the decision support system (DSS) presented in the Figure 5.

The DSS is separated in three different stages, i.e.

- Stage I. Site characterization and description of land-use.
- Stage II. Determination of ecological aspects.

- Stage III. Site-specific tiered assessment (the Triad):
- – Tier 1. Simple screening
- – Tier 2. Refined screening
- – Tier 3. Detailed assessment
- – Tier 4. Final assessment.

Each of these four tiers is based on a weight of evidence (WoE) approach combining three lines of evidence (Chemistry, (eco)Toxicology and Ecology).

3.2.1. Boundaries Of The Dss

The DSS in this chapter is not a full and comprehensive document for managing risk of contaminated land. It focuses strongly on supporting decisions made when considering risk to the terrestrial environment. Therefore it addresses only indirectly the risk to ground water and associated (connected) fresh water systems. Nevertheless information about e.g. reduced bioavailability may be useful when assessing potential risk for leaching of contaminants to ground water or fresh water. Furthermore, it is important to realise that the management of a contaminated site is more than assessing ecological risk. Issues like for example risk for humans, availability and cost of remediation solutions, development plans for the vicinity or the region are equally important.

3.2.2. Stage I — Site Characterisation And Land-Use Definition

The first step in the DSS is to establish what is often referred to as a Conceptual Site Model. It aims at involving as many stakeholders as possible in order to describe site characteristics and to review all available information from the site, e.g. historical information about land-use, investigation of whether the site may be regulated under specific directives, obvious data gaps and urgency for reaction and data collection. The spatial borders of the site should be defined and the current and the future landuse have to be defined. Consultation between administrators, planners and experts therefore has to take place as early as possible in the process.

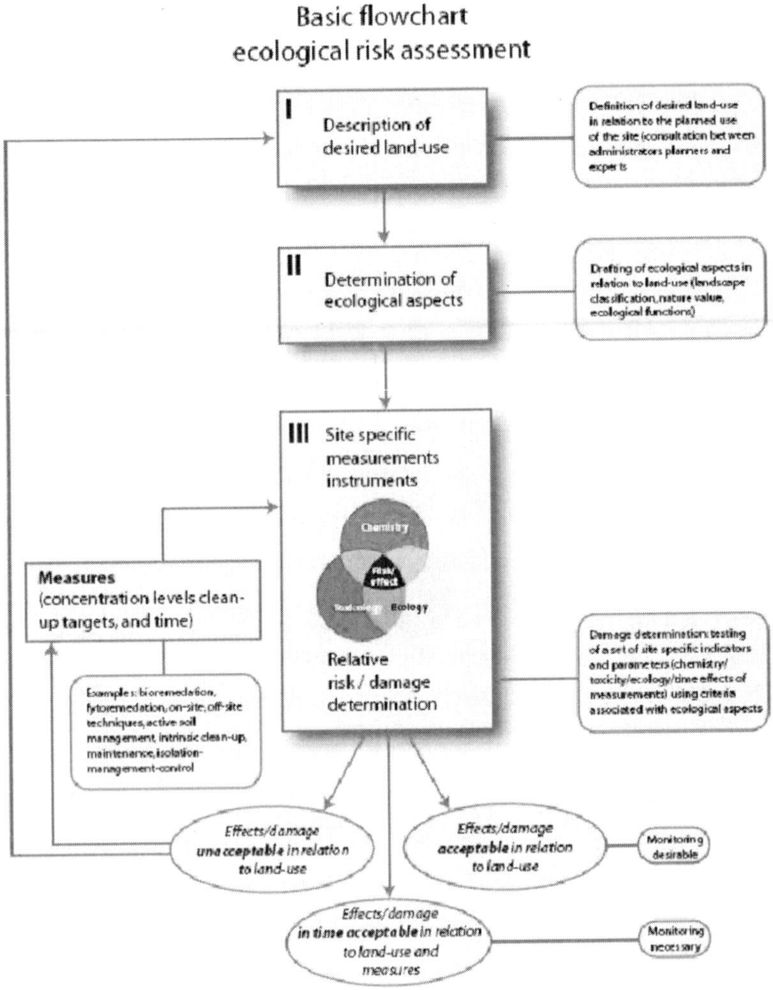

Figure 3. Basic flowchart for ecological risk assessment [90]

Initial requirements in the DSS

An inquiry among all stakeholders should be conducted as one of the first initiatives. The aim should be to collect as much information about soil characteristics as possible.

Defining land-use

One of the first actions to be taken among all stakeholders is to decide which landuse is required for the site, as this will determine the required data collection and testing. Many land-uses may be defined, but generally the four following overall categories of land-use classes are used:

- industrial area (including infrastructure and pavement).
- urban/residential area (including recreational and green areas).
- agricultural area.
- nature area.

When is an ecological risk assessment needed?

Most often a site specific ERA will be initiated only when soil concentrations exceed soil screening levels. However, this may not in itself be a sufficient criterion to go through the entire ERA procedure. Some boundary conditions, based on the present and future type of land-use, the level of contamination and various ecological considerations have to be met in order to rationalize an ERA. The experts and the rest of the stakeholders should answer a number of simple questions in order to conclude whether the required boundary conditions are fulfilled.

3.2.3. Stage Ii — Determination Of Ecological Aspects

At stage II, site-specific ecological features and receptors relating to the land-use defined in Stage I need to be outlined. This includes aspects like key species and life support functions. The potential ecological receptors should be identified in order to determine whether potential source-pathway-receptor linkages can be established. This includes not only ecological receptors directly linked to the site but also those linked indirectly e.g. through leaching of contaminants to connected fresh water systems or (migrating) birds or mammals feeding in the area. In Table 9 some examples are given of land-use and related ecological aspects. This table can be used as a starting point for the selection of ecological aspect. Experts from ecotoxicology and ecology should be involved in the selection of ecological aspects.

3.2.4. Stage Iii — Site Specific Instruments (The Triad)

If after finalising Stage I and Stage II it is still considered that there is a need for a site specific evaluation of ecological risk the process continues to Stage III using the weight of evidence approach described below.

3.2.5. Weight Of Evidence Approaches

In order to deal with conceptual uncertainties in a pragmatic way, it has been proposed to use weight of evidence (WoE) approaches for ERA [90-93]. The rationale is, like in justice, that many independent ways to arrive at one conclusion will provide a stronger evidence for ecological effects, making ERA less uncertain.

In the sediment research area the application of WoE started at an early stage and was called the Sediment Quality Triad. For terrestrial ecosystems WoE approaches and the Triad are still in a developing stage. The Triad approach is based on the simultaneous and integrated deployment of site-specific chemical, toxicological and ecological information in the risk assessment as given in Figure 6. The major assumption is that WoE in three independent disciplines will lead to a more precise answer than an approach, which is solely based on, for example, the concentrations of pollutants at the site. A multidisciplinary approach will help to minimise the number of false positive and false negative conclusions in ERA. It also gives acknowledgement to the fact that ecosystems are too complex to analyse in one-factorial approaches.

- **Chemistry:** The concentration of contaminants in the environment (totals, bioavailable), accumulated in biota, or modelled via food-chains is used for calculation of risks on the basis of toxicity data from the literature.

- **Toxicology:** Bioassays with species across genera are carried out in order to measure the actual toxicity present in environmental samples from the site.

- **Ecology:** Field ecological observations at the contaminated site are compared to the reference site. Deviations from the reference site, which can be plausibly attributed to the contamination levels, are funnelled into the Triad.

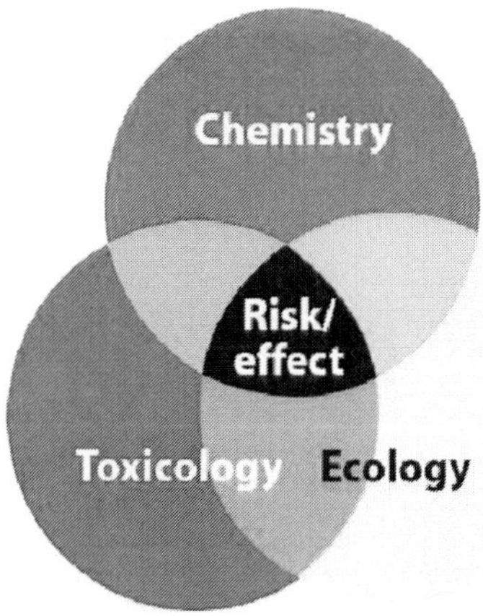

Figure 4. Schematic presentation of the integration of three fields of research according to a Triad [92]

3.3. Using The Triad In Site Specific Assessment Of Contaminated Soil

Triad is a powerful weight of evidence approach originally developed in order to evaluate sediment quality. In the terrestrial compartment less experience is available on the practical use of the Triad. This chapter describes the use of Triad in more detail and gives an insight into some of the important decisions risk assessors have to make when conducting the Triad in practise, e.g. how to scale, weight and integrate the outcome of the various investigations.

The Triad approach exists of three lines of evidence (LoE), the so-called Triad "legs", i.e. chemistry, (eco) toxicology and ecology. The Triad approach includes a tiered system in which each consecutive tier is increasingly fine-tuned to the site-specific situation. In the first tier the research is simple, broad and generic. In later tiers more specific and complex tests and analyses may be used. For each of the LoE in the Triad there are a variety of analyses or tests that can be chosen. Some examples are:

- **Chemistry:** Measurement of total concentrations, bioavailable concentrations, bioaccumulation, etc.

- **Toxicology:** Bioassays (in field and/or in lab), biomarkers etc.

- **Ecology:** Field observations of vegetation, soil fauna, micro-organisms, etc. In Chapter 6, a number of tests or tools that are for suitable for use in each tier are presented for the chemistry, toxicology and ecology LoE.

3.4. Decision Charts In Ecological Risk Assessment Of Contaminated Sites

3.4.1. Flowcharts

This chapter is an attempt to present a decision support system, which can guide risk assessors in their assessment of site-specific ecological risk. A number of site-specific questions need to be answered before a final decision on performing an ecological risk assessment can be made. This chapter introduces a flow chart for ecological risk assessment of contaminated sites. The flowchart is presented as decision trees as shown in Figure 8 together with a more in-depth introduction to the relevant questions that needs to be addressed and answered when performing a site-specific ecological risk assessment.

3.4.2. Decision Making In Era

The assessment of ecological risk is performed stepwise in tiers. Higher tiers represent gradually more and more complex studies, but also more expensive and laborious studies. The full site-specific risk assessment covers four tiers, i.e.

- Simple screening: Tier 1.

- Refined screening: Tier 2.

- Detailed assessment: Tier 3.

- Final assessment: Tier 4.

The main principle in going from a simple screening over a more refined screening to a detailed assessment of the contaminated site is to minimize time and effort. The actual performance of the risk assessment and use of the various tiers may be very site-specific.

Tier 1 — Simple screening

After deciding in the two first stages of the ERA that ecological concern needs special consideration, the risk assessment starts typically with a simple evaluation at the screening level. This is done in order to minimize costs until new information indicates the need for further assessment and more sophisticated studies. Therefore, the tools used in the first screening need not only to be reasonably quick and easy, but also relatively cheap. The tools for use in Tier I are described in more detail in the toolboxes C1, T1 and E1. On the basis of the results of instruments used in Tier 1 it is decided to either stop further assessment or continue to a higher tier.

Tier 2 — Refined screening

Tier 2, still considered being at the screening level, aims at refining the measurement of exposure and at the same time to provide further insight into the toxicological and ecological properties of the contaminated soil. Tier 2 deviate from the conservatism normally associated with the use of total concentration in the risk assessment by taking (rough) estimations of bioavailability into consideration in the chemical LoE. A better screening of the toxicological and ecological properties of the soil compensates for the reduced conservatism in the Chemistry LoE of the Triad. The tools for use in Tier 2 are described in more details in the toolboxes C2, T2 and E2. On the basis of the results in Tier 2 a decision should be made to either stop further assessment or continue to a higher Tier.

Tier 3 — Detailed assessment

The tools in Tier 3 differ from the ones used in Tier 1 and Tier 2 in that they are more laborious, costly and may take longer. On the other hand they are (often) more realistic and/or ecological relevant in order to give a more comprehensive assessment of the ecological risk at the specific site. The stakeholders should beforehand negotiate a minimum set of tests. Is it for example necessary to consider all trophic levels in the toxicological and ecological LoE? Or does the land-use suggest otherwise? Is it necessary (or possible) to estimate the bioavailability of all the substances exceeding their SSL? If not, how are the non-investigated substances dealt with? The tools described for use in Tier 3 are described in more details in the toolboxes C3, T3 and E3. Depending on the results from Tier 3 a decision should be made to either stop further assessment or continue with an even more detailed assessment in Tier 4

Tier 4 — Final assessment

In Tier 4, the aim of the studies is to answer any remaining questions and to decrease existing uncertainties and this may often require more in-depth research. Tools in Tier 4 can be similar to tools of Tier 3, but more focus has to be on site-specific circumstances. For example bioassays should be done with organisms, which normally occur at the site. Furthermore, it may be more relevant to consider ecological effects outside the contaminated area on e.g. predators or herbivores feeding in the area or effects in adjacent fresh water systems. This Tier requires specialised knowledge and experience with ERA, which implies that costs can be high and only a limited number of people may be able to perform the tests. Generally only on a very limited number of site evaluations will include investigations at this level. If the results of Tier 4 still indicate risk there are basically two possible solutions. Accept the risk and leave the contamination or remove (parts of) the contamination.

3.4.3. Screening Tools

3.4.3.1. Triad Based Selection Of Methods

For each of the three Lines of Evidence (LoE) in the Triad various methods or tools are available. In order to facilitate the selection of appropriate tools in the right context, the tools have been compiled in subclasses or toolboxes. Each of these is a collection of tools considered to be potentially useful in the designated tiers and LoE of the Triad, i.e. chemistry, toxicology and ecology. Furthermore, the tools are arranged according to their complexity, price and practicability or in other words depending on whether they are most useful for screening or detailed assessment, i.e.

- Toolbox C1. Chemistry tools for simple screening.
- Toolbox T1. Toxicology tools for simple screening.
- Toolbox E1. Ecology tools for simple screening.
- Toolbox C2. Chemistry tools for refined screening.
- Toolbox T2. Toxicology tools for refined screening.
- Toolbox E2. Ecology tools for refined screening.
- Toolbox C3. Chemistry tools for detailed assessment.

- Toolbox T3. Toxicology tools for detailed assessment.
- Toolbox E3. Ecology tools for detailed assessment.
- Toolbox IV. Various tools for the final (Tier 4) assessments.

Toolbox C1 — Chemistry tools for simple screening

At the very first stage of the ERA process, total concentrations of all relevant chemicals are individually compared to soil screening levels (SSL) in order to evaluate whether there is a need for a site specific assessment of ecological risk. In the current Stage III of the ERA, this first generic evaluation of risk is followed by a more site-specific screening of risk including information from all three lines of evidence in the Triad. In the Chemistry part of the Triad more site-specific information is collected by:

Refining and targeting the comparison of soil concentrations with soil related benchmarks for site-specific purposes.

Incorporation of the accumulative risk of a mixture of contaminants by calculating the toxic pressure (TP) of a mixture and by doing so generating more site-specific insight to the potential ecological impact of a contaminated site. Each of these steps can be done separately or in combination, e.g. the TP can be calculated using existing SSL or using new developed benchmarks based on either NOEC or EC50 values or site-specific benchmarks can be compared to soil concentrations individually. The approach entirely depends on the strategy taken by the stakeholder group and the availability of data.

Toolbox T1 — Toxicology tools for simple screening

The main objective of the selected toxicity tests or bioassay at Tier 1 should be to screen the soil for presence of toxic compounds. This includes toxic degradation products or compounds, which are not routinely included in various national analytical programs for contaminated sites. This Tier is the first screening level of the ERA and the cost in form of manpower and money should hence be relatively low.

Toolbox E1 — Ecology tools for simple screening

Ecological surveys or monitoring studies are generally considered a time consuming effort performed by experts. This is in most cases true, wherefore detailed surveys normally take place in higher tier assessment. However, in order to ensure that also ecological information is collected and used in the

Triad already in the screening phase, it is recommended to perform a limited examination of the site. A survey of the area with special focus on visible changes in e.g. plant cover or presence or absence of specific plants, trees or scrubs may indicate ecological damage, which can be associated to contaminants present at the site.

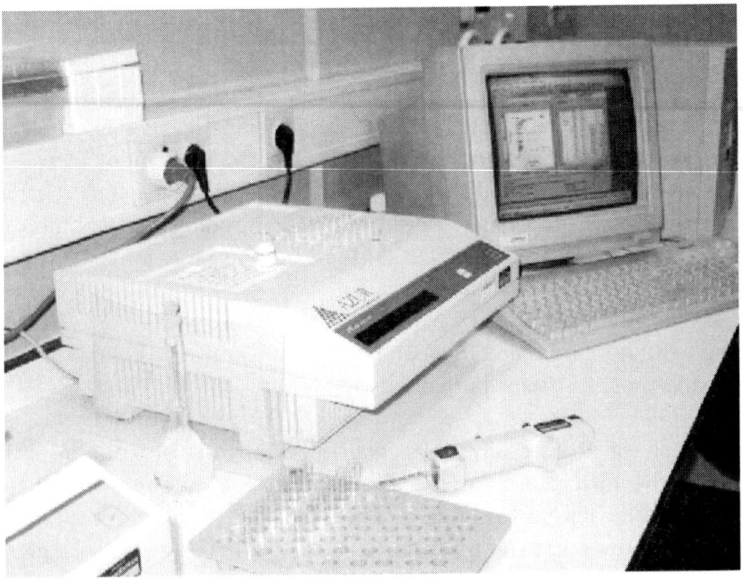

Figure 5. Equipment for measuring luminescence of Vibrio fischeri [94]

If any aerial pictures areavailable from the area these may give valuable information about the plant cover also historically, which may be helpful in identifying parts of the site where the impact may be highest (hot spots). At this stage the conclusion can in most cases only be indicative. Therefore if the results from the other line of evidence may cause any doubt or the survey indicated potential impact, it is recommended to either continue with a more refined screening in Tier 2 or go directly to the detailed assessment in Tier 3.

Toolbox C2 — Chemistry tools for refined screening

Selective solvent extraction

It may be considered useful to adjust the estimate of exposure by taking bioavailability into consideration and hereby deviating from the conservatism normally associated to the use of total concentration in the risk assessment. The principle in this refinement of the ecological risk assessment is to extract a more ecotoxicologically relevant fraction of the contamination than the total concentration. The latter generally tends to overestimate the risk of historically contaminated soils. In this screening phase no attempt is made to estimate the freely dissolved or readily bioavailable concentration of contaminants. Table 10 explains principal studies that employed chemical extractants to evaluate bioavailability.

Figure 6. Simple survey of the site [95]

Outline of principal studies that employed chemical extractants to evaluate bioavailability

Instead the fraction of the contaminants is extracted, which can be directly compared to the existing soil screening levels. This is considered to be a relatively simple and quick method to screen for potential risk of contaminants in a more realistic way than using total concentrations. The extracted concentration (mg kg-1) is compared to the SSL and the result used in the Triad. It is therefore a prerequisite of this comparison that the extractability in the tests (with spiked soils) used for deriving SSL is close to 100% by the methods used. In most short-term tests (< four weeks) it will be reasonable to assume that only little "true" ageing or strong sequestering occurs and hence a majority of the spiked chemicals are still extractable with mild organic solvents. However, for most methods this still has to be fully validated.

Table 5. Outline of principal studies that employed chemical extractants to evaluate bioavailability

Contaminant (Ref)	Solvent	Bioassay	Operation	Comments
Atrazine Phenanthrene (1)	Methanol/water, n-Butanol	Earthworm uptake Degradation	25 ml extractant, 10 g solid. Shaking for 2 h	Methanol/water best predictor for atrazine, n-butanol for phenanthrene.
DDT, DDE, DDD PAH (mixture) (2)	THF Ethanol	Earthworm uptake	15-20 ml extractant, 1 g soil, 10 sec mixing	Good correlation with earthworm accumulation.
Anthracene, Fluoranthene, Pyrene (3)	n-Butanol Propanol Ethyl acetate	Plant retention Earthworm uptake Degradation	25 ml extractant, 1-2 g soil, 5 sec mixing	Reasonable correlation with bioassays.
Phenanthrene Pyrene Chrysene (4)	n-Butanol	Earthworm uptake Degradation	15 ml extractant, 5-10 g soil, Mixing: 5 sec (worm) or 120 sec (degradation)	Applicable for bioavailability prediction.

Organic solvents most frequently used include methanol/water in different ratios, nbutanol, ethanol, propanpol, ethyl acetate and tetrahydrofuran (THF) given in Table 10. The method establishes preferential partitioning of hydrophobic contaminants to the extractant by increasing their solubility in the aqueous phase whilst removing pollutant compounds from soil surfaces establishing equilibrium conditions. No standard protocol has been adopted for mild chemical extractions in relation to bioavailability testing. Common methodology in literature primarily includes a soil sample to which a volume of chemical extractant is added (generally 1 – 10 g soil,15 – 25 ml extractant). This is followed by a period for mixing, e.g. vigorous mixing for10 – 120 seconds or shaking by orbital shakers for up to 2 hours. The extraction studies have mostly involved PAH and insecticides (including DDT, DDE, DDD and atrazine) [93-95]. Studies that haverelated extractability with results from bioassays have generally focused on uptake and accumulation (% taken up by earthworms or plants) and bacterial degradation (%removed). Therefore, since convincing relationships between the chemical and biological tests were found it may indicate a potential for such extraction methods to predict bioavailability.

Toolbox T2 — Toxicology tools for refined screening

In the first simple screening of Tier I focus was on marine bacteria and aquatic/sediment living species. In Tier 2 relatively simple tests with soil dwelling species are used for a more refined screening of the soil samples, i.e. the earthworm survival tests and avoidance tests using soil invertebrates.

Figure 7. Organisms screening in soil [98]

The habitat function of soils is often assessed using the reproduction test with *Eisenia fetida*. The avoidance test with *Eisenia fetida* is a suitable screening test, which is less cost-intensive in terms of duration and workload than the reproduction test, and at the same time (normally) more sensitive than the acute test with the same species.

Toolbox E2 — Ecology tools for refined screening

In Tier 2 the observations from the survey may be expanded by simple on-site assessment of the overall soil functioning or biological activity of the soils. Recommended tools include bait-lamina sticks and simple microbial tests using general endpoints like soil respiration or C/N mineralization rates.

Figure 8. Bait-lamina sticks [99]

The main principle for tests at this level is to be relatively simple and cheap but at the same time to give valuable information whether or not the soil has lost some of its main services. Bait-lamina sticks for example have been demonstrated useful for describing biological activity of the soils in a general matter.

Toolbox C3 — Chemistry tools for detailed assessment

The objective of the tools found in this toolbox is to assess the bioavailable and freely dissolved fraction of pollutants found in pore water of soils from contaminated sites. The methods should (in principle) be able to mimic the fraction of organic pollutants available for uptake in biota. The collection of methods includes various non-depleting and depleting pore water extractions. Very few terrestrial ecotoxicity data are yet expressed as e.g. pore water concentrations. Instead, the outcome of the methodologies in this toolbox is compared with water quality standards.

Toolbox T3 — Toxicology tools for detailed assessment

The objective of the tools found in this toolbox is to evaluate the potential impact of contaminated soils to fauna and plants and hereby the entire ecosystems. Some of the methods use introduced, and not intrinsic, species. The benefit of this is a higher degree of standardisation, as the species used in these bioassays is easy to maintain in laboratory cultures compared to naturally occurring species. The drawback may be that their ecological relevance is less obvious. For example the compost worm*Eisenia fetida* is used as a surrogate to evaluate risk to soil dwelling earthworms. Two sets of bioassays are presented. One for directly assessing potential risk for soil dwelling species, including micro-organisms, plants and soil invertebrates, and one for assessing indirectly risk to aquatic species through e.g. leaching of contaminants. It is often anticipated that soil organisms are exposed to pollutants mainly through uptake from pore water. Therefore it may also be possible to evaluate, or at least to compare or rank, the risk of contaminated soil samples to soil dwelling organisms on the basis of the outcome of the aquatic test using elutriate or pore water. The choice of bioassays depends on a number of variables, e.g.:

- The current and future land-use, i.e. targets of protection.
- The size of the contaminated area.
- The potential for ground water or surface water contamination.
- The need of many simple tests or fewer more complicated tests.

Figure 9. Simple plant tests [91]

Toolbox E3 - Ecology tools for detailed assessment

In this late tier of the Triad, the objective of the activities is community or population response analysis, typically by conducting field surveys. As these studies (most often) are time consuming, costly and dependent on ecologically, taxonomically and statistical expertise they are most frequently done on large-scale sites with a long-term- remediation perspective. In fresh water ecosystem community surveys have been widely used with relative success. The absence of species from places where they would be expected to occur could be a strong identification of unacceptable levels of contaminants. However, this type of studies has only seldom been used for the terrestrial environment. The reasons for this are many. One of the dominants may be the lack of a concentration gradient and obvious "upstream" reference sites at most contaminated areas. No world-wide accepted guideline on how to plan and perform a terrestrial field survey is available and hence no straight-forward and easy-to-follow description can be given. The decision on when, where and how to conduct field surveys depends on a number of issues, e.g. the size of the area, the land-use, the type of contaminants present, time of the year and last but not least the time and money available to perform the study. Nevertheless, a number of general considerations have to be made in the planning phase of a successful field survey. These include (but are not limited to):

- Identify the targets of concern and the species to monitor.
- Elucidate the natural temporal and spatial variation before initiating a field study.
- Use statistical (power) analyses to determine the minimum number of samples or replicates needed to emonstrate the decided difference, e.g. 25% change.
- In order to establish a cause-effect relationship, a number of confounding parameters need to be characterized both at the reference and the test site, e.g. soil type, pH, salinity, hydrology, nutrient- and organic matter content and the presence of other contaminants.

As no single description on how to perform ecological surveys for contaminated sites can be given, some general considerations and useful references for this tier of the ecological risk assessment are given below for:

- Assessing impact in the overall biological activity and organic matter breakdown.
- Assessing impact on the microbial community.

- Assessing impact on the plant community.
- Assessing impact on the invertebrate community.

Reference data from reference sites, reference samples and literature

A crucial factor in a risk assessment is the quality of reference data, because the results of the site-specific ecological measurements or calculations are compared against these data. This is true for as well chemical information (i.e. background levels in that region), toxicological data from bioassays (i.e. site relevant reference soil and well characterized control soil in order to verify the test performance) and ecological field surveys.

Figure 10. Litter bags [94]

The reference soil should in principle resemble the contaminated soil in all relevant parameters, e.g. texture, pH, organic matter, water-holding capacity, nutrient content. In practice, these ideal spots are difficult to find. If there is no or inadequate reference information, effects can only be determined in relative terms by comparison with other sites. This is usually adequate for determining the degree of urgency and/or the need for remediation. Reference data can be obtained by including reference sites (preferably more than one) in the sampling scheme, including reference measurements in the experimental set-up, or by obtaining reference data from the literature or by expert based judgment.

Higher tier assessment of the impact on biological activity and organic matter breakdown

In addition to the general information about biological activity in soils generated in Tier 2 from the bait-lamina test, other, slightly more laborious, tests may give additional information about the overall biological activity in soil, e.g. wheat straw decomposition (litter bag test) and cotton strip degradation. A review paper from Van Gestel et al. (2003) concluded that while the bait-lamina gave the best reflection of the biological activity of soil animals, e.g. earthworms, springtails and enchytraeids, the litter bag test and the cotton strip test are more indicative of the microbial activity in the soil [96]. Knacker et al. (2003) reviewed the usefulness of five different litter decomposition tests and concluded that the litter-bag test had distinct advantages over the others [97]. All of these simple tests only give insight into the overall activity in soils and the breakdown of organic material. They are hence most suitable on their own in cases of land-use with low sensitivity, e.g. industrial land. For land-uses where structural endpoints, e.g. biodiversity or specific species, are the target of protection other endpoint(s) should be monitored as well.

Higher tier assessment of the impact on the microbial community

The number of microorganisms, especially bacteria, in soil is extremely large. They differ widely in their function and sensitivity to chemicals. Besides more classical (and simple) measures of the microbial community like total bacterial biomass, the number of colony forming units and substance induced respiration rate (SIR), more advanced methods for assessing the impact of contamination on soil microorganisms have recently been made available.

This include microbial "fingerprinting" like phospholipid fatty acid analysis (PLFA) and community-level physiological-profiling (CLPP) based on metabolic response using BIOLOG plate systems, and the use of pollution induced community tolerance (PICT).

Higher tier assessment of the impact on the plant community

Plants interact dynamically with the physical and chemical characteristics of soils. Soil types and site characteristics, therefore, greatly influence the occurrence of plants and their total aboveground biomass (also called, Net Primary Production, i.e. NPP) within given climatic conditions and human management. Vegetation cover is an important indicator of soil quality and a diverse plant community is normally a good indication of essential soil functions such as the decomposition process, the mineralisation rate, and the occurrence of soil dwelling animals associated to (fresh) organic matter. Vascular plants are easy to sample. They are immobile and hence associated to soil contamination (and airborne pollution). Plant community parameters like plant cover, aboveground plant biomass, plant shoot/ root ratio, species diversity and the binary occurrence (presence/absence) of specific indicator species like metal-tolerant species may be used successfully in ERA.

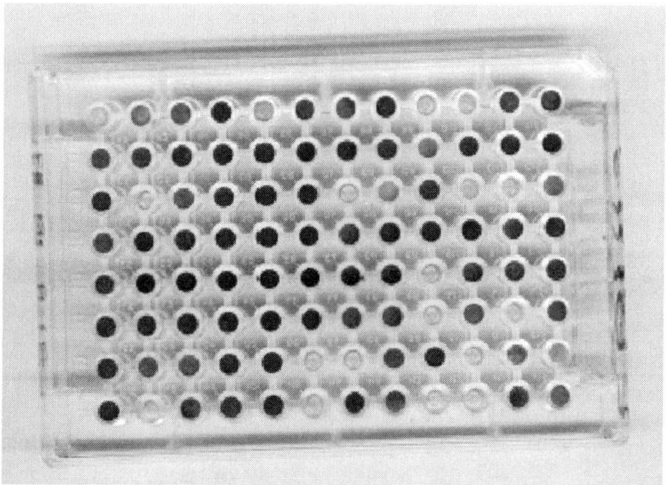

Figure 11. Microbial community [96]

Higher tier assessment of the impact on the soil invertebrate community

Survey of soil biota in order to evaluate the effect of various sources of pollution on soil communities on historically contaminated sites have not yet been used on a larger scale by e.g. consultants. However, numerous (monitoring) studies by various research groups can be found in the open literature. Methods of surveying include:

- Collection of soil samples followed by extraction in the laboratory.
- Extraction or collection of organisms in the field, e.g. by hand-sorting or by the application of mustard or formalin.

Figure 12. Plant survey [95]

- Trapping (surface dwelling) animals by the use of e.g. pit-falls. Monitoring species includes earthworms, snails, oribatid mites, nematodes, springtails, ants, ground-living beetles and spiders. Most of the studies have been done on metal contaminated sites (see references below). A substantial amount of work has been put into the challenge of developing a soil invertebrate system for evaluating risk of pollutants. The only soil invertebrate system that is used on a regularly basis in the context of ecological risk assessment of contaminated soils is most likely the nematode Maturity Index (MI). The system is based on the evidence, that rapid colonising species dominate nematode communities in disturbed ecosystems. In the Netherlands experience with surveys of soil invertebrates from the monitoring programme Biological Indicator for Soil Quality (BISQ) has also been used in ERA.

Toolbox for tests in Tier IV

The final assessment in the ERA process is not likely to be initiated for many contaminated sites. The choice of additional tests or monitoring at this level of the ERA is bound to be very site-specific and hence an issue for negotiation between stakeholders and experts. Accumulation in biota is included in this toolbox as the internal concentration in biota is believed, at least to some extend, to reflect uptake and then bioavailability. An alternative in this final tier could also be to model uptake in biota provided sufficient data is available [98].

Figure 13. Soil fauna sampling [98]

Figure 14. Detailed field survey [93]

4. SOIL MANAGEMENT PROGRAM (REMEDIATION)

4.1. Reporting Norms and Standards For Contaminated Land

The aim is to investigate whether the land has been contaminated, and if contamination has occurred whether the contamination presents a significant risk of harm. The commonly encountered international practice consists of three distinct reporting phases. It progresses from Phase 1 desktop and site walkover assessments with limited investigation and testing to a Phase 2 detailed invasive investigation and testing for site characterisation to a comprehensive Phase 3 report with an evaluation of remediation objectives and a proposed remediation plan, supported by control and monitoring measures for the activities.

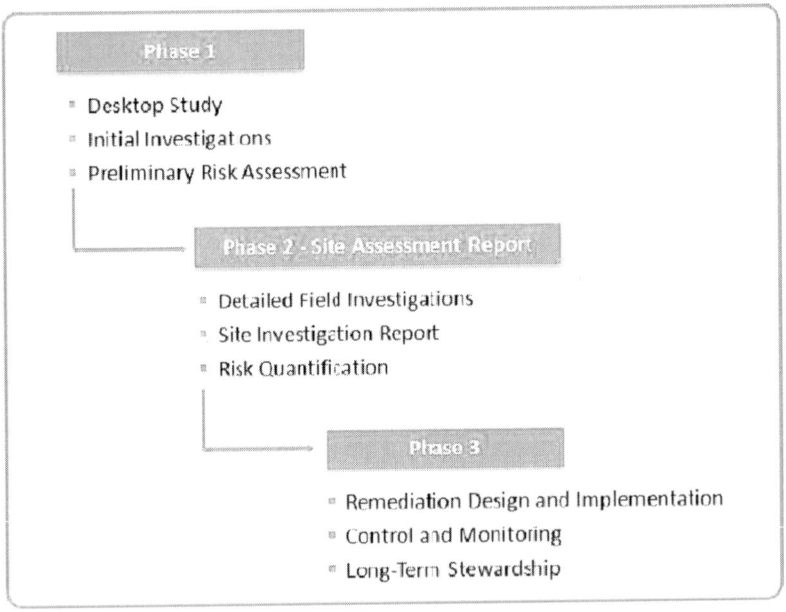

Figure 15. A phase approach for the assessment and remediation of contaminated land

The reporting system requires norms and standards of practice to be strictly applied, but also must retain flexibility to allow for decisions on the contaminated status of sites to be made in the most beneficial manner (considering ecological, social and economic aspects) also taking into account timeframes. In some cases, urgent priority works may require that the phased approach to reporting has to move forward in a concurrent single report.

4.2. Requirements For Preliminary Site Assessment Phase 1

A preliminary site assessment must consider the following elements

1 Site description-location and size

2 Nature and extent of contamination, contaminants of concern and historic activities that may be sources of contamination. List all present and past activities at the site that involves storage and production, use, treatment or disposal of hazardous material that could contaminate the site.

3 Describe current condition of the site and contents and the results of any previous assessment report

4 Local topography and geology, drainage, surface cover, vegetation.

5 Status of ground water, approximate depth to water table

6 Proximity to surface water

7 Proximity to drinking water supplies

8 Annual rainfall and flood potential

9 Land and water use for the nearby areas and site

10 Any other regulations as Regulated by the ministry

All data may not be available, or data may vary in terms of uncertainty, it is thus important to recognize gaps in the knowledge base and to decide whether additional data must be obtained on the site characterisation. This may trigger the commencement of Phase 2 Investigations. The Phase 1 report must make clear recommendations on the status of the contamination risk posed by the site. If a complete site history clearly demonstrates that the site activities do not pose a contamination threat then no further investigation is warranted and the site should be recommended as suitable for reuse. In most cases it is likely that some level of preliminary investigation will be required to provide the level of certainty required to enable property re-development or transfer. A limited investigation of certain subsurface activities, for example, underground storage tanks, would be necessary to obtain a waiver on the contamination status of a site at a Phase 1 level of reporting.

4.3. Approval Requirements

If soil contaminants are found at concentrations that exceed the applicable standards specified in the approval, the approval holder is required to implement a soil management program.The program must first address source control to stop on-going contaminant releases. After the sources of contamination have been stopped, further assessment and delineation of the contaminated area may be necessary. When the extent of contamination is understood, remediation objectives, as described below, must be adopted for the area. Once remediation objectives have been agreed to, appropriate treatment or containment technologies can be chosen and the management plan finalized.

4.4. Developing Remediation Objectives

Remediation objectives may be developed in a variety of ways ranging from generic guidelines to site-specific risk assessment. Generic guidelines are numerical concentration limits that are applicable under a variety of site conditions. When neither Environmental Protection nor the agency has a guideline for a particular substance, four options are available. First, the remediation objective may be based on the ambient background concentration for the site. Second, the guideline development protocol may be applied. Third, a remediation objective may be adopted from another jurisdiction if the proponent can show that the remediation objective is consistent with the environmental protection goals of the approval. Finally, a remediation objective may be developed by the proponent using site-specific risk assessment procedures. A risk assessment for an approved facility will focus on human health concerns but fundamental ecological concerns must also be addressed. Site-specific risk assessment is a means of quantifying the likelihood that soil contamination will have a harmful effect under conditions found at a specific site. The essential components of human health and ecological risk assessments are similar; however, ecological risk assessments tend to be more complex than those for human health because a wide variety of receptors may have to be considered. Very briefly, a site-specific risk assessment may be described as consisting of the following steps or components as shown in Figure 20.

Problem formulation involves developing a conceptual model of the possible contaminant effects on receptors at the site. The conceptual model describes contaminant distribution and concentration in relation to the receptors and their patterns of activity on the site. The *exposure assessment* describes the pathways by which soil contaminants may be taken up by the receptor. This information is combined with receptor characteristics in order to estimate the contaminant uptake. The *toxicity assessment*describes the adverse effects that the contaminants may cause and the dose at which these effects occur.

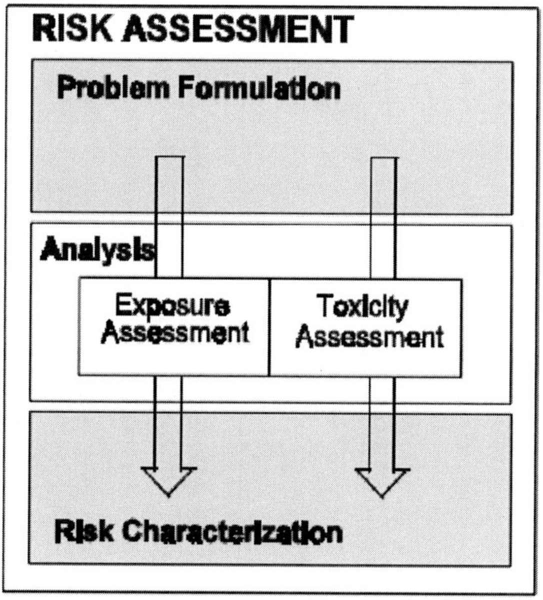

Figure 16. Risk Assessment Framework

The final step, *risk characterization*, compares this dose with the uptake rate (estimated during the exposure assessment) and determines whether or not an adverse effect is likely to occur. Risk assessment procedures can also be used to back calculate a contaminant concentration in soil at which no adverse effects are expected. At times, it may not be possible to remediate to a level compatible with industrial land-use objectives. In such cases, the approval holder must ensure that the contaminants are contained and receptor exposure does not occur. Formal risk assessment procedures are necessary to meet these requirements. Typically, an engineered containment system is required, the performance of which must be confirmed through periodic inspections and monitoring. The approval holder bears responsibility for the design, construction, assessment and maintenance of the risk management system, and any necessary emergency response should the system fail.

4.5. Enforcement Policy

Enforcement is generally the last step in a regulatory process that aims to first prevent potential problems before they arise and to resolve existing problems in a cooperative manner. When discussions between the approval holder and the Department fail to prevent or resolve a contravention of an approval or the Act, a number of abatement and enforcement tools are available to the Department including:

- environmental protection orders;
- warning letters;
- tickets;
- administrative penalties;
- enforcement orders;
- prosecutions;
- court orders; and
- cancellation of approvals or certificates.

Due to various factors, including the high cost involved in remediation interventions, it is essential that a holistic and tiered, risk-based approach be adopted that is founded on international best practice, to address remediation in a uniform manner across the country. This is relevant irrespective of the sector of occurrence to safeguard both human health and the natural environment. The Framework is based on a review of international practice in the developed countries of the world and the emergence of remediation policy from developing countries, and an assessment of alternative approaches and methodologies that may find application in the development of a remediation framework.

5. CONCLUSION

Soil pollution is a result of many activities and experiments done by mankind which end up contaminating the soil. Industrial wastes such as harmful gases and chemicals, agricultural pesticides, fertilizers and insecticides are the most common causes of soil pollution. The others are ignorance towards soil management and related systems, unfavourable and harmful irrigation practices, improper septic system and management and maintenance of the same, leakages from sanitary sewage. There is urgent need for a tiered approach in ecological risk assessment of contaminated soils. Generic soil screening levels are needed as a first tier. Higher tiers of ecological risk assessment should, however, contain some kind of site-specific assessment. It is furthermore important to organize the various studies in a framework or decision support system that is transparent and useful for all stakeholders. A weight of evidence approach may be an obvious choice to deal with these uncertainties. The TRIAD approach, which incorporates and categorizes information in a triangle – chemistry, toxicology, and ecology – is an appropriate tool for handling conceptual uncertainties. Several remedies to these shortcomings have been proposed. Regarding ecotoxicity direct testing would allow for a major improvement in risk estimates. As to human health risks: including biological availability in risk estimates, more use of up to date knowledge about exposure routes, dose-effect relations and combination effects, and biomonitoring of effects are options for improvement.

ACKNOWLEDGEMENTS

Authors greatly acknowledge the University Malaya Research Grant (RG257-13AFR). Moreover, the comments of two anonymous reviewers are gratefully acknowledged.

REFERENCES

1. Okrent D. 1999. On intergenerational equity and its clash with intragenerational equity and on the need for policies to guide the regulation of disposal of wastes and other activities posing very long time risks. Risk Analysis 19: 877-901.

2. Belluck, D.A., Benjamin, S.L., Baveye, P., Sampson, J., Johnson, B. 2003. Widespread arsenic contamination of soils in residential areas and public

spaces: an emerging regulatory or medical crisis? International Journal of Toxicology 22: 109-128.

3. Richardson, G.M., Bright, D.A., Dodd, M. 2006. Do current standards of practice in Canada measure what is relevant to human exposure at contaminated sites? II: oral bioaccessibility of contaminants in soil. Human and Ecological Risk Assessment 12: 606-618.

4. Van Zorge, J. A. 1996. Exposure to mixtures of chemical substances: is there a need for regulations? Food and Chemical Toxicology 34, 1033-1036.

5. Swartjes, F.A. 1999. Risk-based assessment of soil and groundwater quality in the Netherlands: standards and remediation urgency. Risk Analysis 19:1235-1248.

6. Eikelboom, R.T., Ruwiel E., Goumans, J.J.M. 2001. The building materials decree: an example of a Dutch regulation based on the potential impact of materials on the environment. Waste Management 21: 295-302.

7. Crane, M. and Giddings, J.M. 2004. 'Ecologically acceptable concentrations' when assessing the environmental risks of pesticides under European Directive 91 414/EEC. Human and Ecological Risk Assessment 10: 733-747.

8. Nathanail, P., McCaffrey, C., Earl, N., Forster, N.D., Gillett, A.G., Ogden, R. 2005. A deterministic method for deriving site-specific human health assessment criteria for contaminants in soil. Human and Ecological Risk Assessment 11: 389-410.

9. Tarazona, J.V., Fernandez, M.D., Vega, M.M. 2005. Regulation of contaminated soils in Spain. Journal of Soil and Sediments 5:121-124.

10. Evans, J., Wood, G., Miller, A. 2006. The risk assessment-policy gap: An example from the UK contaminated land regime. Environment International 32: 1066-1071.

11. Huinink, J.T.M. 1998. Soil quality requirements of use in urban environments. Soil and Tillage Research 47: 157-162.

12. Urzelai, A., Vega, M., Angulo, E. 2000. Deriving ecological risk-based soil quality values in the Basque Country. Science of the Total Environment 247: 279-284.

13. Apitz, S.E. 2008. Is risk based, sustainable sediment management consistent with European policy. Journal of Soils and Sediments 8: 461-466.

14. Provoost, J., Cornelis, C., Swartjes, F. 2006. Comparison of soil clean-up standards fort race elements between countries: why do they differ? Journal of Soil and Sediments 6: 173-181.

15. Provoost, J., Reijnders, L., Swartjes, F., Bronders, J., Carlon, C., D'Allessandro, M., Cornelis, C. 2008. Parameters causing variation between soil screening values and the effect of harmonization. Journal of Soils and Sediments 8: 298-311.

16. Toccalino, P.L. and Norman, J.E. 2006. Health-based screening levels to evaluate US Geological Survey groundwater quality data. Risk Analysis 26: 1339-1348.

17. Patterson, M.M., Cohen, E., Prommer, H., Thomas, D.G., Rhodes, S., McKinley, A.I. 2007. Origin of mixed brominated ethane groundwater plume: contaminant degradation pathways and reactions. Environmental Science & Technology 41: 1352-138.

18. Leon Paumen, M. 2008. Invertebrate life cycle responses to PAC exposure. PhD thesis. Amsterdam: University of Amsterdam.

19. Nawrot, T., Plusquin, M., Hogervorst, J., Roels, H.A., Celis, H., Thijs, L., Vangronsveld, J., Van Hecke, E., Staessen, J. 2006. Environmental exposure to cadmium and risk of cancer: a prospective population-based study. The Lancet Oncology 7: 119-126.

20. Simcox, N.J., Fenske, R.A., Wolz, S.A., Lee, I.W., Kalman, D.A. 1995. Pesticides in household dust and soil: exposure pathways for children of agricultural families. Environmental Health Perspectives 103: 1126-1134.

21. Laidlaw, M.A.S., Mielke, H.W., Filippelli, G.M., Johnson, D.L., Gonzales, C.R. 2005. Seasonality and children's blood lead levels: developing a predictive model using climate variables and blood data for Indianapolis, Indiana, Syracuse, New York and New Orleans Louisiana (USA). Environmental Health Perspectives 113: 793-800.

22. Laidlaw, M.A.S., Mielke, H.W., Filippelli, G.M., Johnson, D.L. 2006. Blood lead in children: Laidlaw et al. respond. Environmental Health Perspectives 114: A 19.

23. Rothenberg, S.J. and Rothenberg, J.C. 2005. Testing the dose-response specification in epidemiology: public health and policy consequences for lead. Environmental Health Perspectives 113: 1190-1195.

24. Akesson, A., Lundh, T., Vahter, M., Bjellerup, P., Lidfeldt, J., Nerbrand, C., Samisoe, G., Strömberg, U., Skerfving, S. 2005. Tubular and glomerular kidney effects in Swedish women with low environmental cadmium exposure. Environmental Health Perspectives 113: 1627-1631.

25. Nawrot, T.S., van Hecke, E., Thijs, L., Richart, T., Kuznetsova, T., Jin, Y., Vangronsveld, J., Roels, H.A., Staessen, J.A. 2008. Cadmium-related mortality and long-term secular trends in the cadmium body burden of an environmentally exposed population. Environmental Health Perspectives 116: 1620-1628.

26. Lamphear, R.P., Dietrich, K., Auinger, P., Cox, C. 2000. Cognitive defects associated with blood lead concentrations, 10 microgram/dl in United States children and adolescents. Public Health Reports 115: 521-529.

27. Wang, C.L., Chuang, H.J., Ho, C.K., Vang, C.Y. 2002. Relationship between blood lead concentrations and learning achievement among primary schoolchildren in Taiwan. Environmental Research A 89: 167-181.

28. Wu, T., Buck, G.M., Mendola, P. 2003. Blood levels and sexual maturation in US girls. Environmental Health Perspectives 111: 737-741.

29. Von Storch, H., Costa-Cabral, M., Hagner, C., Freser, F., Pacyna, Y., Pacyna, E., Kolb, S. 2003. Four decades of gasoline lead emissions and control policies in Europe: a retroactive assessment. Science of the Total Environment 311: 151-176.

30. Mielke, H.W., Gonzales, C.R., Powell, E., Jartun, M., Mielke Jr., P.W.2007. Nonlinear association between soil lead and blood lead of children in Metropolitan New Orleans, Louisiana: 2000-2005. Science of the Total Environment 388: 43-53.

31. Provoost, J., Cornelis, C., Swartjes, F. 2006. Comparison of soil clean-up standards for trace elements between countries: why do they differ? Journal of Soils and Sediments 6: 173–181.

32. Miranda, M.L., Kim, D., Overstreet-Galeano, M.A., Paul, C.J., Hull, A.P., Morgan, S.P. 2007. The relationship between early childhood blood lead levels and performance on end-of-grade tests. Environmental Health Perspectives 115: 1242-1247.

33. Patterson, B. M., Cohen, E., Prommer, H., Thomas, D. G., Rhodes, S., McKinley, A. J. 2007. Origin of a mixed brominated ethene groundwater

plume: contaminant degradation pathways and reactions. Environmental Science & Technology 41: 1352–1358.

34. Huel, G., Sahuquillo, J., Debotte, G., Oury, J., Takser, L. 2008. Hair mercury negatively correlates with calcium pump activity in human term newborns and their mothers at delivery. Environmental Health Perspectives 116: 263-267.

35. Nawrot, T., Plusquin, M., Hogervorst, J., et al., 2006. Environmental exposure to cadmium and risk of cancer: a prospective population-based study. The Lancet Oncology 7: 119–126.

36. Laidlaw, M. A. S., Mielke, H. W., Filippelli, G. M., Johnson, D. L. 2006. Blood lead in children: Laidlaw et al. respond. Environmental Health Perspectives 114: A19-A23.

37. Akesson, A., Lundh, T., Vahter, M., et al., 2005. Tubular and glomerular kidney effects in Swedish women with low environmental cadmium exposure. Environmental Health Perspectives 113: 1627–1631.

38. Mergler, D., Anderson, H.A., Chan, L.H.M., Mahaffey, K.R., Murray, M., Sakamoto, M., Stern, H.A. 2007. Methylmercury exposure and health effects in humans: a worldwide concern. Ambio 36: 3-11.

39. ten Tusscher, G.W. 2002. Later childhood effects of perinatal exposure to background levels of dioxins in the Netherlands. PhD thesis. Amsterdam: University of Amsterdam.

40. J. A. Van Zorge, "Exposure to mixtures of chemical substances: is there a need for regulations?"Food and Chemical Toxicology, vol. 34, no. 11-12, pp. 1033–1036, 1996.

41. Filser, J., Koehler, H., Ruf, A., Rombke, J., Prinzing, A., Schaefer, M. 2008. Ecological theory meets soil ecotoxicology: challenge and chance. Basic and Applied Ecology 9: 346-355.

42. Van Gestel, C.A.M. 2008. Physico-chemical; and biological parameters determine metal bioavailability in soils. Science of the Total Environment 406: 385-395.

43. Sokolik, G.A., Ovsiannikova, S.V., Ivanova, T.G., Leinova, S.L. 2004. Soil-plant transfer of plutonium and americium in contaminated regions of Belarus after the Chernobyl catastrophe. Environment International 30: 939-947.

44. Zhao, C., Ren, J., Xue, C., Lin, E., 2005. Study on the relationship between soil selenium and plant selenium uptake. Plant and Soil 277: 197-206.

45. Liste, H., Prutz, I. 2006. Plant performance, dioxygenase-expressing rhizosphere bacteria, and biodegradation of weathered hydrocarbons in contaminated soil. Chemosphere 62: 1411-1420.

46. van der Geest, H, Leon Paumen, M. 2008. Dynamics of metal availability and toxicity in historically polluted floodplain sediments. Science of the Total Environment 406: 419-425.

47. Brouwer, A., Ahlborg, U.G., van den Berg, M., Birnbaum, L.S. 1995. Functional aspects of developmental toxicity of polyhalogenated hydrocarbons in experimental animals and human infants. European Journal of Pharmacology 293: 1-40.

48. Rutgers, M., Tuinstra, J., Spijker, J., Mesman, M., Wintersen, A., Posthuma, L. 2008. Risico's voor het ecosysteem in stap twee van het saneringscriterium [Ecosystem risks in step 2 of the remediation criterion]. Bilthoven (the Netherlands): RIVM.

49. Li, J., Zhou, B., Liu, Y., Yang, Q., Cai, W. 2008. Influence of coexisting contaminants on bisphenol A sorption and desorption in soil. Journal of Hazardous Materials 151: 389-393.

50. Gennings, C., Carter Jr, W.H., Casey, M., Moser, V., Carchman, R., Simmons, J.E. 2004. Analysis of functional effects of five pesticides using a ray design. Environmental Toxicology and Pharmacology 14: 115-126.

51. Hayes, T.B, Case, P., Chui, S., Chung, D., Haeffele, C., Haston, K., Lee, M., Mai, V.P., Marjuoa, Y., Parker, J., Tsui, M. 2006. Pesticide mixtures, endocrine disruption and amphibian declines: are we underestimating the impact. Environmental Health Perspectives 114, Supplement 1: 40-50.

52. Perry, M.J., Venners, S.A., Barr, D.B., Xu, X. 2007. Environmental pyrethoid and organophosphorous insecticide exposures and sperm concentration. Reproductive Toxicology 23: 113-118.

53. Laetz, C.A., Baldwin, D.H., Clllier, T.K., Hebert, V., Stark, J.D., Scholz, N.L. 2009. The synergistic toxicity of pesticide mixtures: implications for risk assessment and the conservation of the endangered Pacific salmon. Environmental Health Perspectives 117: 348-353.

54. Roos, P. H., Tschirbs, S., Pfeifer, F., et al., 2004. Risk potentials for humans of original and remediated PAH-contaminated soils: application of biomarkers of effect. Toxicology 205: 181–194.

55. Xiao, R. Y., Wang, Z., Wang, C. X., Yu, G., Zhu, Y. G. 2006. Genotoxic risk identification of soil contamination at a major industrialized city in northeast China by a combination of in vitro and in vivo bioassays. Environmental Science & Technology 40: 6170–6175.

56. Lee, D. H., Lim, J. S., Song, K., Boo, Y., Jacobs Jr., D. R. 2006. Graded associations of blood lead and urinary cadmium concentrations with oxidative-stress-related markers in the U.S. population: results from the third National Health and Nutrition Examination Survey. Environmental Health Perspectives 114: 350–354.

57. De Zwart, D. and Posthuma L. 2005. Complex mixture toxicity for single and multiple species: proposed methodologies. Environmental Toxicology and Chemistry 24: 2665-2676.

58. Altenburger, R., Walter, H., Grote, M. 2004. What contributes to the combined effect of a complex mixture? Environmental Science & Technology 38: 6353-6362.

59. Filser, J., Wittmann, R., Lang A. 2000. Response types in Collembola towards copper in the microenvironment. Environmental Pollution 107: 71-78.

60. Drexler, J.W. and Brattin, J.W. 2007. An in vitro procedure for estimation of lead relative bioavailability: with validation. Human and Ecological Risk Assessment 13: 383-401.

61. O'Halloran, K. 2006. Toxicological considerations of contaminants in the terrestrial environment for ecological risk assessment. Human and Ecological Risk Assessment 12: 74-83.

62. Römbke, J. 2006. Tools and techniques for the assessment of ecotoxicological impacts of contaminants in the terrestrial environment. Human and Ecological Risk Assessment 12: 84-101.

63. Roos, P.H., Tschirbs, S., Pfeifer, F., Welge, P., Hack, A., Wilhelm, M., Bolt, H.M. 2004. Risk potentials for humans of original and remediated PAH-contaminated soils: application of biomarkers of effect. Toxicology 205: 181-194.

64. Xiao, R., Wang, Z., Wang, C., Yu, G., Zhu, Y. 2006. Genotoxic risk identification of soil contamination at a major industrialized city in north east China by a combination of in vitro and in vivo bioassays. Environmental Science & Technology 40. 6170-6175.

65. Lee, D., Lim, J., Song, K., Boo, Y., Jacobs, D.R. Jr. 2006. Graded associations of blood lead and urinary cadmium concentrations with oxidative- stress-related markers in the US population: Results of the Third National Health and Nutrition Examination Survey. Environmental Health Perspectives 114: 350-354.

66. Neumann, H.G. 1996. Toxic equivalence factors, problems and limitations. Food and Chemical Toxicology 34: 1045-1051.

67. Dobrev, I.D., Andersen, M.E., Yang, R.S.H. 2002. In silico toxicology: simulating interaction thresholds for human exposure to mixtures of trichloroethylene, tetrachloroethylene and 1,1,1 trichloroethane. Environmental Health Perspectives 110: 1031-1039.

68. Silva, E., Rajapakse, N., Kortenkamp, A. 2002. Something from 'nothing'- eight weak estrogenic chemicals combined at concentrations below NOECs produce significant mixture effects. Environmental Science & Technology 36: 1751-1756.

69. Richter, M. and Escher, B. 2005. Mixture toxicity of reactive chemicals by using two bacterial growth assays as indicators of protein and DNA damage. Environmental Science & Technology 39: 8753-8761.

70. Caeiro, S., Costa, M. H., Ramos, T. B. 2005. Assessing heavy metal contamination in Sado Estuary sediment: An index analysis approach. Ecological Indicators 5: 151–169.

71. Håkanson, L. 1980. An ecological risk index for aquatic pollution control: A sedimentological approach. Water Research 14: 975–1001.

72. Long, E. R., MacDonald, D. D. 1998. Perspective: Recommended uses of empirically derived, sediment quality guidelines for marine and estuarine ecosystems. Human and Ecological Risk Assessment 4: 1019–1039.

73. Pekey, H., Karakaş, D., Ayberk, S., et al., 2004. Ecological risk assessment using trace elements from surface sediments of İzmit Bay (Northeastern Marmara Sea) Turkey. Marine Pollution Bulletin 48: 946–953.

74. Roach, A. C. 2005. Assessment of metals in sediments from Lake Macquarie, New South Wales, Australia, using normalization models and sediment quality guidelines

75. Sutherland, R. A. 2000. Bed sediment-associated trace metals in an urban stream, Oahu, Hawaii. Environmental Geology 39: 611–627

76. Bhattacharya, A., Routh, J., Jacks, G., et al., 2006. Environmental assessment of abandoned mine tailings in Adak, Västerbotten District (Northern Sweden). Applied Geochemistry 21: 1760–1780

77. Usero, J., González-Regalado, E., Gracia, I. 1996. Trace metals in the Bivalve Mollusc Chamelea Gallina from the Atlantic Coast of Southern Spain. Marine Pollution Bulletin 32: 305–310.

78. Cheng, J. L., Shi, Z., Zhu, Y. W. 2007. Assessment and mapping of environmental quality in agricultural soils of Zhejiang Province, China. Journal of Environmental Sciences, 19: 50–54

79. Reeves, W.R., Barhoumi, R., Burghardt, R.C., Lemke, S.J., Mayura, K., McDonald, T.J., Phillips, T.D., Donelly, K.C.2001. Evaluation of methods for predicting the toxicity of polyaromatic hydrocarbon mixtures. Environmental Science & Technology 35: 1630-1636.

80. El Masri, H., Mumtaz, M.M., Yushak, M.L. 2004. Application of physiologically-based pharmacokinetic modeling to investigate the toxicological interaction between chlorpyrifos and parathion in rat. Environmental Toxicology and Pharmacology 16: 57 71.

81. Kortenkamp, A. and Altenburger, R. 1999. Approaches to assessing combination effects of estrogenic environmental pollutants. Science of the Total Environment 333: 131-140.

82. Sumpter, J.P. and Johnson, A.C. 2005. Lessons from endocrine disruption and their application to other issues concerning trace organics in the aquatic environment. Environmental Science & Technology 39: 4321-4332.

83. Brian, J.V, Harris, C.A., Scholze, M., Backhaus, T., Booy, P., Lamoree, M., Pojana, G., Jonkers, N., Runnalis, T., Bonfä, A., Marcomini, A., Sumpter, J. 2005. Accurate prediction of the response of freshwater fish to a mixture of estrogenic chemical. Environmental Health Perspectives 113: 721-728.

84. Wölfe, D. 1997. Interactions between 2,3,7,8 TCDD and PCs as tumor promoters: limitations of TEFs. Teratogenesis, Carcinogenesis and Mutagenesis 17: 217-224.

85. Marchant, C.A. 1996. Prediction of rodent carcinogenicity using the DEREK system for 30 chemicals currently being tested by the National Toxicology Program. Environmental Health Perspectives 104: 1065-1073.

86. Verhaar, H.J.M., Morroni, J.R., Reardon, K.F., Hays, S,M., Gaver, D.P., Carpenter, R.L., Yang, R.S.H. 1997. A proposed approach to study the toxicology of complex mixtures of petroleum products. Environmental Health Perspectives 195 (Supplement 1): 179-195.

87. Stevenson, C.N., Macmanus-Spencer, L.A., Luckenbach, T., Luthy, R.C., Epel, D. 2006. New perspectives on perfluorochemical ecotoxicology: inhibition and induction of an efflux transporter in the marine mussel Mytilus californianus. Environmental Science & Technology 40: 5580-5585.

88. Bailer, A.J., Hughes, M.R., See, K., Noble, R., Schaefer, R. 2002. A pooled response strategy for combining multiple lines of evidence to quantitatively estimate impact. Human and Ecological Risk Assessment 8:1597-1612..

89. Chapman, P.M., McDonald, B.G., Lawrence, G.S. 2002. Weight-of-evidence issues and frameworks for sediment quality (and other) assessments. Human and Ecological Risk Assessment 8:1489-1515.

90. Rutgers, M., den Besten, P. 2005. Approaches to legislation in a global context.The Netherlands perspective – soil and sediment. In:Toxicity Testing Environmental. Thompson C, Wahhia K, Loibner AP (eds). Blackwell, CRC Press, Oxford, pp 269–289.

91. Alexander, M. 1995. How toxic are toxic chemicals in soil? Environmental Science and Technology. 29:2713–2717.

92. Alexander, M. 2000. Aging, bioavailability, and overestimation of risk from environmental pollutants. Environmental Science and Technology. 34:4259–4265.

93. Callahan, C. A., Menzie, C. A., Burmaster, D. E.,Wilborn, D. C., Ernst, T. 1991 On-site methods for assssesing chemical impact on the soil environment using earthworms: a case study at the Baird and McGuire superfund site, Holbrook, Massachusetts. Environmental Toxicology and Chemistry. 10:817–826.

94. Attrill, M. J., Depledge, M. H., 1997. Community and population indicators of ecosystem health: Targeting links between levels of biological organisation. Aquatic Toxicology 38:183–197.

95. Keddy, C., Greene, J. C., Bonnell, M. A. 1994. A review of whole organism bioassays for assessing the quality of soil, freshwater sediment and freshwater in Canada. Scientific series no. 198. Environment Canada, Ottawa, Ontario, Canada.

96. Van Gestel, C. A. M., Kruidenier, M., Berg, M. P. 2003. Suitability of wheat straw decomposition, cotton strip degradation and bait-lamina feeding tests to determine soil invertebrate activity. Biology and Fertility of Soils 37:115-123.

97. Knacker, T., Förster, B., Römbke, J., Frampton, G.K. 2003. Assessing the effects of plant protection products on organic matter breakdown in arable fields – litter decomposition test systems. Soil Biology and Biochemistry 35:1269-1287.

98. Jager, T., Fleuren, R.H.L.J., Hogendoorn, E.A., De Korte, G. 2003. Elucidating the routes of exposure for organic chemicals in the earthworm, Eisenia andrei (Oligochaeta). Environmental Science and Technology 37:3399-3404.

Chapter 2

Metabolomics for Soil Contamination Assessment

Maria C. Hernandez-Soriano[1] and Jose C. Jimenez-Lopez[2]

[1] School of Agriculture and Food Sciences, The University of Queensland, St Lucia QLD, Australia

[2] The UWA Institute of Agriculture, The University of Western Australia, Crawley, Perth WA, Australia

1. INTRODUCTION

The evaluation of biological responses to assess and predict the impact of environmental changes in ecosystems functioning is receiving increasing attention, and current research focuses on overcoming concerns about the specificity of biomarkers (Amiard-Triquet et al., 2012). Generally, biomarkers are chemicals, metabolites, susceptibility characteristics, or physiological changes that relate to the exposure of an organism to a chemical. Accordingly, a selected biomarker, i.e. biological response, can be linked to a specific environmental exposure, being representative of the health status of the ecosystem studied. The identification of biomarker profiles has been possible upon the development of metabolomics. Those profiles allow the genuine identification of the relevant biological response/s associated to a particular exposure while the assessment of a single biomarker could only estimate the potential response of the ecosystem to a particular pollutant.

Metabolomics is the generic name assigned to a scientific field that addresses the characterization of low molecular weight organic metabolites released by living organisms in response to environmental stimuli. Morrison et al. (2007) provided an extended definition "the application of metabolomics to the investigation of both free-living organisms obtained directly from the natural environment (whether studied in that environment or transferred to a laboratory for further experimentation) and of organisms reared under laboratory conditions (whether studied in the laboratory or transferred to the environment for further experimentation), where any laboratory experiments specifically serve to mimic scenarios encountered in the natural environment".

The methodological approach of metabolomics relies on a comprehensive analysis of the set of metabolites or "metabolome" produced in response to particular environmental stimuli. Accordingly, the metabolome is the pool of metabolites, small molecules, within a cell, tissue, organ, biological fluid, or entire organism (Miller, 2007). Exposure of an organism to an external stressor will result in changes at the level of the metabolome (Ankley et al., 2006; van Ravenzwaay et al., 2007), and such changes may constitute a highly sensitive indicator of an external stress. Therefore, metabolomics has potential as a sensitive and rapid technique that can elucidate the relationships between metabolite levels and an external stressor, such as contaminant exposure, nutritional deficit or a disease.

The main advantage of metabolomics over traditional research is to overcome the bias associated to the assessment of predefined metabolites (Singh, 2006). Among the diverse applications of omic profiling methods to the environmental sciences, ecotoxigenomics addresses the response of organisms to pollutants based on the different sensitivity of species to toxicants (Spurgeon et al., 2008). Currently, the implementation of metabolomics in ecological risk assessment is still at an early stage, mostly applied as a screening tool to assess the potential toxic effects of pollutants or to determine the mode of action (MOA) of a toxicant. Otherwise, application of metabolomics for environmental monitoring allows the study of a large variety of species from relatively uncontrolled environments.

Bundy et al. (2009) highlight the challenge of identifying a large number of metabolites and the necessity of creating metabolite databases specifically dedicated to environmental issues. Multivariate statistical analysis has proved highly effective for metabolite identification. Thus, principal component analysis (PCA) is used to identify differences between metabolic profiles of organisms exposed to organic or inorganic pollutants (Jones et al., 2014; Kwon et al., 2012; Lankadurai et al., 2011). Besides, association between the metabolic profile and biological factors evaluated as markers for exposure to pollutants can be modelled by partial least squares (PLS) regression analysis (Ellis et al., 2012).

The implementation of metabolomics for the assessment of soil contamination is nevertheless at an early stage (Viant, 2009). A basic screening of published research in the web of science returns circa 100 items for the search "soil pollution-metabolomics", with a significant launch in 2011 (Figure 1), reduced to 21 records when the search is narrowed with the term "biomarkers", published in the period 2007-2013. However, emerging regulatory challenges demand the advance of toxicity testing. Toxicogenomics tools have been presented as an advanced from the current methodologies used for regulatory decision making in ecotoxicology, which entirely rely on whole animal exposures and adverse effects on survival, growth, and reproduction (Ankley et al., 2006). From the acquisition of reproducible metabolic profiles as response to the presence of specific pollutants in soil (Jones et al., 2008) to the application of metabolomics techniques to the study of the response of the entire community of a soil to factors such as pollution and climate change (Jones et al., 2014), the implementation of metabolomics in ecotoxicology is a sound answer to the current needs of society and the environment (van Ravenzwaay et al., 2012).

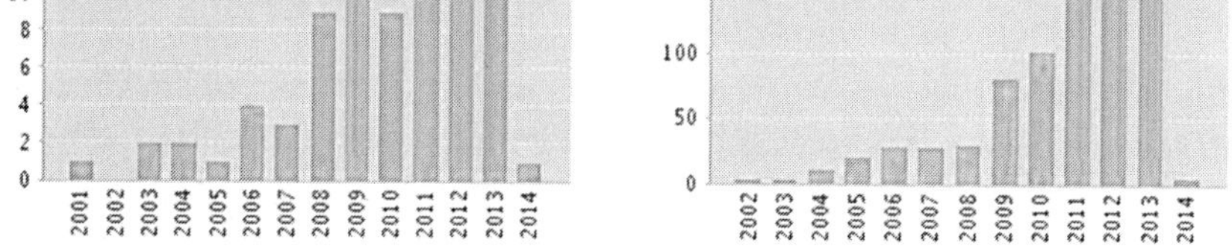

Figure 1. Citation report for the search "soil pollution-metabolomics" as obtained from Thomson Reuters (January 2014).

During the last decade a number of general revisions about the application of metabolomics in environmental health assessment have been published (Bundy et al., 2009; Miller, 2007; Snape et al., 2004; van Ravenzwaay et al., 2007; Viant et al., 2003). The present review specifically summarizes the most significant research concerning implementation of metabolomics in soil contamination assessment. The main objectives of this revision are i) to provide a systematized outline for the application of metabolomics in risk assessment of soil contamination, ii) to provide a rapid guide to the methodological approaches currently optimized and iii) to unify and simplify the knowledge currently available in the topic to provide an accessible tool for further advance in the implementation of metabolomics in risk assessment.

2. METHODOLOGICAL APPROACHES

2.1. Metabolites Isolation

Generally, metabolites are extracted from intact organisms (occasionally from selected relevant tissues) that have been exposed to the studied toxicant by moderate chemical extraction (Baylay et al., 2012;Yuk et al., 2010). The organisms commonly selected for toxicity testing are earthworms (Table 1), particularly the genus *Eisenia*, which are a classic model organism for toxicity assays (Sanchez-Hernandez, 2006; Van Gestel et al., 1992; van Gestel et al., 1989) and have been since long included in official guidelines (OECD, 1984, 2004). Earthworms ingest large amounts of soil and uptake a significant amount of contaminant through the skin. Therefore they are continuously exposed to contaminants. Extractions performed with methanol-chloroform (Baylay et al., 2012) or phosphate buffer solution (Yuk et al., 2010) on pulverized or lyophilised organisms are described to extract the maximum number of metabolites while allowing the performance of reliable analyses.

The isolated extracts usually might not require further sample treatment prior to analysis, which minimizes the introduction of artefacts but also facilitates the development of low cost, rapid methodologies.

2.2. Metabolites Determination: Chromatography, Spectroscopy And Spectrometry

The leading analytical techniques in metabolomics for soil contamination assessment are proton nuclear magnetic resonance spectroscopy (^{1}H NMR) and gas chromatography–mass spectrometry (GC-MS), as thoroughly reported in Table 1, both allowing the identification of compounds at molecular level in the analysed substances. Several authors have also implemented high pressure liquid chromatography (HPLC) and ultra high pressure liquid chromatography (UPLC) coupled with mass spectrometry detector (MS) for the assessment of biological responses to soil contamination with heavy metals (Hédiji et al., 2010; Hughes et al., 2009). Overall, these analytical techniques allow the determination and identification of the metabolites that foremost represent the metabolic alterations related to the toxic effects of organic or inorganic contaminants in soil.

Table 1 Metabolic responses of test organism following exposure to selected environmental contaminants in contact tests.

Technique	Organic toxicant	Animal model tested	Biomarkers of contaminant exposure	Reference
^{1}H NMR GC/MS	2-fluro-4-methylaniline	*E. veneta*	[D][2-hexyl-5-ethyl-3-furansulfonate, maltose], [I]inosine monophosphate	(Bundy et al., 2002)
^{1}H NMR GC/MS	3-trifluromethyaniline	*E. veneta*	[I]lactate	(Lenz et al., 2005)
^{1}H NMR GC/MS	3-trifluromethylaniline	*E. veneta*	[I][Ala, Gly, Asn, glucose, citrate, succinate]	(Warne et al., 2000)
^{1}H NMR GC/MS	3-fluro-4-nitrophenol	*E. veneta*	[D][acetate, malonate], [I][succinate, trimethylamine-N-oxide]	(Bundy et al., 2001)
^{1}H NMR GC/MS	3,5-difluroaniline	*E. veneta*	[D]2-hexyl-5-ethyl-3-furansulfonate, [I]inosine monophosphate	(Bundy et al., 2002)
^{1}H NMR GC/MS	4-fluroaniline	*E. veneta*	[D][Maltose, hexyl-5-ethyl-3-furansulfonate]	(Bundy et al., 2002)
^{1}H NMR	Aroclor 1254	*E. fetida*	No significant changes	(Fitzpatrick et al., 1992)
^{1}H NMR	Caffeine	*E. fetida*	[I]fumarate	(McKelvie et al., 2011)
^{1}H NMR	Carbamazephine	*E. fetida*	[D][Fumarate, Glu, Val, [D]Leu]	McKelvie et al. (2011)
^{1}H NMR	Carbaryl	*E. fetida*	[D][Phe, Tyr, Lys, Ala, Val, Leu]	(Heimbach, 1988) (Edwards and Bater, 1992)
^{1}H NMR	Chlorpyrifos	*E. fetida*	No significant changes	(Yu et al., 2006)
^{1}H NMR	Chlorpyrifos	*L. rubellus*	[I]fumarate	(Baylay et al., 2012)

^{1}H NMR spectroscopy, GC–MS	Chlorpyrifos	*C. elegans*	^{I}[Ala, betaine, ornithine], ^{D}[choline, Glu, Gly, Ile, lactate, n-buterate, taurine]	(Baylay et al., 2012)
^{1}H NMR	Dimethyl phthalate	*E. fetida*	^{D}[Phenylalanine, Alanine, Leucine, Valine]	(Drewes and Vining, 1984)
^{1}H NMR GC/MS	DTT and Endosulfan	*E. fetida*	^{I}[maltose, Ala, Leu]	(McKelvie et al., 2009)
1-D & 2-D ^{1}H NMR spectroscopy	Endosulfan	*E. fetida*	Significant fluctuations in glutamine/GABAeglutamate cycle metabolites and spermidine	(Yuk et al., 2013)
1-D & 2-D ^{1}H NMR spectroscopy	Endosulfan Sulfate	*E. fetida*	Significant fluctuations in glutamine/GABAeglutamate cycle metabolites and spermidine	(Yuk et al., 2013)
^{1}H NMR	Estrone	*E. fetida*	^{D}[Adenine, Glu]	McKelvie et al. (2011)
^{1}H NMR GC-MS analysis	Imidacloprid/ Thiacloprid	*L. rubellus*	^{D}Toxicological endpoints (survival, weight loss, and reproduction)	Baylay et al. (2012)
^{1}H NMR	Naphthalene	*E. fetida*	^{D}[Ala, Leu, Val, Lys]	(Brown et al., 2009)
^{1}H NMR	Nonylphenol	*E. fetida*	^{D} [adenine, Glu]	McKelvie et al 2011
^{1}H NMR	Polybrominated diphenyl ethers (PBDE) 209	*E. fetida*	^{D}Maltose, ^{I}[Lys, Glu]	McKelvie et al. (2011)
^{1}H NMR	Perfluorooctanoic acid	*E. fetida*	^{I}[succinate, HEFS, Glu], ^{D}[leu, Val, LyS, Phe, Arg, maltose, ATP]	(Lankadurai et al., 2012)
^{1}H NMR	Perfluorooctane sulfonate	*E. fetida*	^{I}[succinate, HEFS, Glu], ^{D}[leu, Val, Lys, Phe, Arg, maltose, ATP]	(Lankadurai et al., 2012)
^{1}H NMR	PHA	*E. fetida*	Leu, Val, Ala, Lys and maltose changed in response to PAH exposure.	(Brown et al., 2009)
^{1}H NMR	Phenanthrene	*E. fetida*	^{I}[Ala, betaine, Scyllo- and myo-inositol, cholesterol, phosphatidylcholine], ^{D}Glu	(Lankadurai et al., 2011) Brown et al. (2010)

^{1}H NMR	Polychlorinated biphenyl	*E. fetida*	^{I}ATP	McKelvie et al. (2011) (Whitfield Åslund et al., 2011)
^{1}H NMR	Poly brominated diphenyl ethers	*E. fetida*	^{I}[lys, Glu], ^{D}maltose	McKelvie et al. (2011)
^{1}H NMR GC/MS	Pyrene	*L. rubellus*	^{D}[lactate, tetradecanoic acid, hexadecanoic acid, octadecanoic acid], ^{I}[Ala, Leu, Val, Ile, Lys, Tyr, methionine]	(Jones et al., 2008)
1-D & 2-D ^{1}H NMR spectroscopy	Rifluralin	*E. fetida*	^{I}[Ala, Gly, maltose, ATP]	Yuk et al. (2011)
^{1}H NMR	Thiacloprid	*L. rubellus*	No significant changes were reported	Baylay et al. (2012)
1-D & 2-D ^{1}H NMR spectroscopy	Trifluralin	*E. fetida*	^{I}[Ala, Gly, ATP], ^{D}maltose	Yuk et al. (2011)
^{1}H NMR spectroscopy and UPLC-MS	Cd	*C. elegans*	^{I}phytochelatins, ^{D}cystathionine	(Hughes et al., 2009)
^{1}H NMR, HPLC-PDA	Cd	*S. lycopersicum*	^{I}choline, ^{D}[glucose, citrate, malate, glutamine, asparagine, Phe, Tyr, Val, Ile, trigonelline]	(Hédiji et al., 2010)
^{1}H NMR, GC-MS analysis	Chlorpyrifos + Ni	*L. rubellus*	^{I}[Phe, glucose, malate, arachidonic acid, fumarate, Lys, Tyr, monosaccharides, monophosphorylated form of inositol, succinate, uracil, ethanolamine, Pro, putrescine], ^{D}myo-inositol,	(Baylay et al., 2012)
^{1}H NMR spectroscopy, GC–MS	Chlorpyrifos + Ni	*C. elegans*	^{I}[Ala, creatine, His, lactate, betaine, carnosine], ^{D}[choline, Gly, Ile, leu, lys, Val]	(Jones et al., 2012)
^{1}H NMR	Cu (II)	*L. rubellus*	^{I}His	(Gibb et al., 1997)

^{1}H NMR spectroscopy, GC–MS	Ni	*C. elegans*	^{I}[Asn, choline, Gln, lactate, succinate], ^{D}[Ala, Ile, Gly, Val]	Jones et al. (2012)
^{1}H NMR	Pb, Zn	*S. salsa*	i) Exposure to Pb - ^{D}Tyr. ii) Exposure to Zn - ^{I}[Val, Ile, Leu, Thr, Ala, Asn, Phe, fosfocoline], ^{D}[acetate, Gly, glucose, fumarate and ferulate]. iii) Exposure to Pb & Zn - ^{I}[Ala, Asn, Tyr, Phe], ^{D}[acetate, succinate, Asp, malonate, fructose, glucose, fumarate and ferulate]	(Wu et al., 2013)
^{1}H NMR	Mixed metals (Cadmium, copper, lead and zinc)	*L. rubellus*	^{I}[maltose, His]	(Bundy et al., 2004)
^{1}H NMR	Tellurite	*P. pseudoalcaligenes*	^{I}[Thr, Leu, Tyr, betaine, Ser, Lys, Ile, Ala, Arg, Val, glutathione, adenosine], ^{D}[Lactate, Gly]	(Tremaroli et al., 2009)
^{1}H NMR	Titanium dioxide	*E. fetida*	^{I}[Phe, Tyr, Lys, Glu, Ala, lactate, Val, Leu], ^{D}maltose	(Whitfield Aslund et al., 2012)

D = decrease; I = increase

2.3. Metabolomics Data Analysis

The general approach to data analysis in metabolomics can be summarized in three main stages: explorative, supervised and biological interpretation (Smilde et al., 2010). The explorative phase aims to find groups, clusters and outliers in metabolites and samples studied while the supervised discriminates two or more groups to make predictive models and to find biomarkers (Amiard-Triquet et al., 2012; Dallinger-Marianne, 2000; van Ravenzwaay et al., 2007). Multivariate methods are currently preferred, although univariate and semi-univariate methods have been commonly used for selecting biomarkers. For instance, the lysosomal system was identified as a particular target for the toxic effects of pollutants in soil organisms. However, it is nonspecific as a marker and only included in a suite of biomarkers among diverse soil invertebrate species can provide the necessary specificity for risk assessment purposes (Kammenga et al., 2000). Finally, the biological interpretation seeks the links between metabolome data and underlying metabolic networks through metabolite set enrichment, pathway analysis and metabolic network inference (Trygg et al., 2006). Thus, finding metabolite relationships is essential to determine comprehensive and meaningful metabolic changes as biological response to environmental stimuli (Ellis et al., 2012; Morrison et al., 2007). Accordingly, such extensive evaluation of the impact of pollutants in the metabolism of target organisms is the approach that can add value to the assessment of soil health and viability of soil organisms undergoing stress from pollution.

3. METABOLOMICS BIOINFORMATICS

Information processing by bioinformatics tools and computational biology methods has become essential for solving complex biological problems in genomics, proteomics, and metabolomics. Understanding "omics" data requires both common statistical and computational based methods due to the multi-dimensional and complexity level of the data.

Data-analytical methods for the study of biological systems as developed in the field of computational biology provide a suit of indispensable tools to survey the outcome of metabolomics studies. First, computational biology allows a fast screening of the large biological and chemical data sets generated (Shulaev, 2006), and therefore the identification of the most relevant metabolites, i.e. compounds specifically representative of the metabolic changes in the model system following exposure to different concentrations of organic and inorganic toxicants. As a result of the large number of variables (metabolites) studied, metabolomics studies encompass a significant statistical power for the systematic detection of biological responses to environmental changes (van Ravenzwaay et al., 2012). Second, the mathematical models developed in computational biology allow the identification of relationships between the external stimuli and the metabolic response (Zhang et al., 2010). Third, the implementation of computational algorithms to structural biology makes possible to discover the structure-function of new macromolecular compounds, the functional enzymatic conversion and changes in their activity, as well as their molecular interaction and relationship with others compounds in the pathways where they are involved (Jimenez-Lopez et al., 2013). Moreover, it is possible to detect patterns in such biological responses and establish significant dose-response relationships. Besides, pattern recognition reduces the metabolomics data from hundreds of variables to two or three components that are orthogonal to each other. Overall, this advance of computational biology has been possible due to three significant technological breakthroughs: high-information-content data streams, novel bio-statistical methods, and the computational power to analyse these data.

Data processing and statistical analyses are commonly performed using multivariate (typically a principal component analysis (PCA) and (or) partial least squares (PLS) regression analysis) and univariate (t-test) analyses (Brown et al., 2010; Jones et al., 2014; McKelvie et al., 2011; Yuk et al., 2013). These analyses are performed in combination with the quantification and identification of the metabolites. Subsequently, biological interpretation of the data is necessary for understanding the link between the external stimulus and the metabolic response of the organisms.

Principal component analysis is the most widely used multivariate statistical approach in metabolomics, used to explain the overall variability in a data set via a a set of uncorrelated variables called principal components (PCs), which are linear combinations of the original variables (Trygg et al., 2006). The organization of samples in PCA scores plots is based on the similarities between their metabolic profiles. Thus, PCA allows for dimensional reduction of the data into a low dimensional plane, such as PC1 versus PC2. The scores plot (e.g., PC1 versus PC2) allows for a visual examination of the relationship between the samples based on their metabolic profiles. In a 1-D PCA loadings plot, the contribution (or weight) of each metabolite to the discrimination of the sample classes along one component is represented by the intensity of the metabolite peak. In the 2-D PCA loadings plot discrimination is performed by selecting the points that are scattered further away from the tight cluster of points found near the origin.

Other widely used multivariate statistical tools in metabolomics are PLS regression analysis and PLS discriminant analysis (PLS-DA). Both PLS-regression and PLS-DA are methods for samples classification, with pre-defined variables added to maximize the separation between the sample classes and to construct predictive models. The predefined variables for PLS- regression are measurable quantities such as the contaminant exposure concentration. Validation methods such as the leave-one-out cross validation are used to test the robustness of the models generated by PLS-regression, PLS-DA, OPLS, and OPLS-DA (Whitfield Åslund et al., 2011).

Although metabolomics studies mostly use multivariate statistics, univariate statistical analyses can contribute to the information gained from a study. Thus, t-tests can be used to assess the significance of the separation between the controls and stressed organisms in PCA and PLS-DA scores plots. Also, t tests can be used to determine which metabolites in the ^{1}H NMR spectra of the treatment class increased or decreased significantly relative to the controls.

4. BIOMARKERS

The somewhat secondary significance of biological responses for soil contamination assessment was customarily associated to the limitation of biomarkers as measurable responses to contaminants, which classically could only provide an indication of exposure to contaminants in soil (Sanchez-Hernandez, 2006). The development of metabolomics, considered an "emerging field" as late as mid-2010, has provided the tools for the determination of multiple biomarkers across different levels of biological organization, and therefore a better assessment of the ecological consequences of contamination. Since the creation of the first metabolomics web database, METLIN (Smith et al., 2005), 60,000 metabolites has been incorporated, a rapid development closely related to the evolution of mass spectrometry instrumentation and data analysis tools. Currently, the number of databases and metabolites registered is continuously increasing. Table 2 summarizes some of the most relevant databases operative and the corresponding website is also indicated. Further information on metabolomics databases can be obtained from the metabolomics society (http://www.metabolomicssociety.org). For instance, ChemSpider is an aggregated database of organic molecules containing more than 20 million compounds from many different providers. At present the database contains information from such diverse sources as a marine natural products database, ACD-Labs chemical databases, the EPA's DSSTox databases and from a series of chemical vendors. It has extensive search utilities and most compounds have a large number of calculated physico-chemical property values.

One of the goals in bioinformatics is to establish automated and efficient ways to integrate large, biological datasets from multiple sources. This objective is challenging because data sources are heterogeneous in terms of their functions, structures, data access methods and dissemination formats. In addition, the enormous quantity of information produced by "omics" is handled via computers that systematically analyze and store the accumulating sequence, structure and function data. Databases are essential in metabolomics because they provide a rapid and specific tool to identify the compounds isolated from an organism exposed to a particular environmental challenge. Thus, the KNApSAcK package provides tool for analysing datasets of mass spectra as well as for retrieving information on metabolites by entering the name of a metabolite, the name of an organism, molecular weight or molecular formula. A list of metabolites that are associated to a taxonomic class can be obtained by search with the taxonomic name, from which information of individual metabolites can be retrieved. The NIST Chemistry WebBook provides access to chemical and physical property data for chemical species. The data provided in the site are from collections maintained by the NIST Standard Reference Data Program and outside contributors. Data in the

NIST Chemistry WebBook can be found by direct searches for chemical species or indirect searches based on related data. Specific databases are also being developed, such as LIPID MAPS, currently the largest database of lipid molecular structures. Otherwise, SetupX combines mass spectrometric and biological metadata, which is a step forward in the organization of information generated by metabolomics analysis.

Table 2. Selected metabolomic databases.

METLIN	http://metlin.scripps.edu/index.php
LIPID MAPS	http://www.lipidmaps.org/
KEGG	http://www.genome.jp/kegg/pathway.html
ChemSpider	http://www.chemspider.com/
SetupX	http://fiehnlab.ucdavis.edu/projects/binbase_setupx
KNApSAcK	http://kanaya.naist.jp/KNApSAcK/
NIST	http://webbook.nist.gov/chemistry/
MassBank	http://www.massbank.jp/
HMP	http://www.hmdb.ca/
IIMDB	http://metabolomics.pharm.uconn.edu/iimdb/

[3] - Source http://www.metabolomicssociety.org/databases.

Metabolomic databases are thus accompanied by accurate description of the biological study design and accompanying metadata reporting on the laboratory workflow from sample preparation to data processing.

Currently, standard analyses focus on the determination of amino acids, mono- and disaccharides, lipids/fatty acids, short chain fatty acids and small phenolics. Accordingly, it is possible to already launch the standardization of metabolomics analysis. For instance, the Northwest Metabolomics Research Center (University of Washington) has established a relevant list of target compounds to evaluate biological responses to changes in the environment. The list of compounds is summarized inTable 3.

The information of metabolites and metabolic pathways has been obtained from the website of Kyoto Encyclopedia of Genes and Genomes (Kegg, http://www.genome.jp/kegg/). According to the research results summarized in Table 1, the implementation of metabolomics in the assessment of soil contamination indicates that contaminants in soil affect several of the major metabolic pathways in living organisms (Table 3), including glycolysis, trycarboxylic acids cycle and amino acids metabolism. Moreover, data analysis indicates an overall reduction in the production of the associated metabolites. For instance, the interference in amino acids specialized pathways results in a decreased synthesis of purine and pyrimidine nucleotides (Brown et al., 2010; McKelvie et al., 2011). These nucleotides are essential for the production of the energy (ATP molecules) that drive most of the enzymatic reactions in living organisms, but also protein synthesis is consequently hampered, which explain the negative effect in processes such as antioxidant activity.

Table 3. Summary of metabolites and metabolic pathways representative of biological responses to environmental stimuli.

Metabolic Pathways	Number of Metabolites
Alanine, aspartate and glutamate metabolism	15
Arginine and proline metabolism	23
Butanoate metabolism	18
Citrate cycle (TCA cycle)	11
Cysteine and methionine metabolism	14
Fatty acid metabolism	3
Glutathione metabolism	14
Glycine, serine and threonine metabolism	21
Glycolysis / Gluconeogenesis	16
Histidine metabolism	13
Lysine biosynthesis	7
Lysine degradation	6
Nitrogen metabolism	9
Oxidative phosphorylation	6
Pentose phosphate pathway	10
Phenylalanine metabolism	10
Phenylalanine, tyrosine and tryptophan biosynthesis	8
Purine metabolism	30
Pyrimidine metabolism	30
Pyruvate metabolism	10
Synthesis and degradation of ketone bodies	4
Tryptophan metabolism	15
Tyrosine metabolism	18
Valine, leucine and isoleucine biosynthesis	11
Valine, leucine and isoleucine degradation	5

Another emerging group of biomarkers, as highlighted in several studies, are lipids (Rochfort et al., 2009; Sanchez-Hernandez, 2006). Rochfort et al., (2009) indicate that lipophilic extracts can be used in field based metabolomics experiments to investigate different treatment effects on earthworms. Lipid metabolism is highly sensitive to environmental contaminants (Vega-López et al., 2013), with increasing production of lipoprotein vesicle and lipid peroxidation rate during early stages of the biological response to the presence of a toxicant (Lankadurai et al., 2011). Relatedly, earthworm esterases has been proposed as biomarkers for pesticide contamination in soil (Sanchez-Hernandez, 2010). Esterases are directly involved in the natural tolerance of earthworms to pesticides, and can therefore be used as specific biomarkers, but furthermore, their characterization by metabolomics approach might help to select the appropriate earthworm species for regulatory toxicity testing. Overall, the increasing specificity of the research performed in ecotoxigenomics will allow a realistic and meaningful incorporation of biological responses in ecological risk assessment.

5. OXIDATIVE STRESS IN CONTAMINATED SOIL

The induction of the oxidative stress response by the presence of toxic compounds in the environment is a primary mechanisms of defence, although prolonged exposure to contaminants is likely to overwhelm this short-term defence (Regoli et al., 2002).

Metabolites such as proline possibly detoxify the ROS under stress in vivo (Smirnoff, 1993). Exposure of plants to both redox active, for example, Cu and Hg, and other metals, for example, Cd and Zn, induces the generation of free radicals that leads to oxidative stress. This represents one of the major causes of toxicity particularly due to redox metals. The cells are equipped with an elaborate network of antioxidative enzymes and low molecular weight metabolites which mitigate the oxidative stress. Proline scavenges different free radicals in certain in vitro generation and detection systems.

Proline quenches ROS and reactive nitrogen species (RNS), which relieves the oxidative burden from the glutathione system. Moreover, polyamines also have an antioxidative role by quenching the accumulation of $O_2^{\cdot-}$ probably through inhibition of NADPH oxidase (Paschalidis and Roubelakis-Angelakis, 2005). This may facilitate phytochelatin synthesis and enhance metal tolerance (Siripornadulsil et al., 2002).

Overall, oxidative defence response to toxicity or other environmental stress involves the generation of oxygenated metabolites from exposed organisms and activation/inhibition of the production of antioxidants enzymes and metabolites such as glutathione. The depletion of antioxidants for prolonged exposures might result in the decrease of the response effectiveness and eventual imbalance between generation and elimination of reactive oxygen species. Depletion of glutathione appears to be a major mechanism in short-term heavy metal toxicity (Schutzendubel and Polle, 2002). In accordance with this hypothesis, a good correlation between glutathione contents and tolerance index was observed with 10 pea genotypes differing in Cd sensitivity (Metwally et al., 2005). High GSH concentrations in hyperaccumulator *T. Goesingense* coincided with high constitutive activity of serine acetyl transferase (SAT); SAT catalyses the acetylation of L-Ser to OAS which in turn provides the carbon skeleton for Cys biosynthesis. Elevated GSH levels in *T. Goesingense* also coincided with the ability both to hyperaccumulate Ni and to resist its damaging oxidation effects.

The significance of glutathione and the metal-induced phytochelatins (PCs) in heavy metal tolerance has been studied intensely (Rauser, 1995). However, PCs are important for detoxification of only a limited set of metals such as Cd^{2+}, Cu^{2+} and AsO_2^{2-} while Zn^{2+} and Ni^{2+} are poor inducers of PCs and exhibit low binding affinity. Most other metals lack significant binding.

Evaluation of metabolites related to oxidative response constitutes a relevant group of target compounds for risk assessment. Although oxidative response to soil contamination has been classically addressed in plants, the study of this response in soil microorganisms is already being introduced in ecotoxicology as a fundamental part of the biological response of soil microorganisms to soil contamination (Boer et al., 2013; Tremaroli et al., 2009). Accordingly, Boer et al. (2013) describe the attenuation of the oxidative response for springtails in laboratory tests, which constitutes and early detection of soil pollution, and standardized test have been developed.

6. METABOLITES RELATED TO SOIL CONTAMINATION WITH ORGANIC COMPOUNDS

The importance of the identification of biomarkers and metabolic pathways specifically related to soils contamination with a particular pollutant or group of pollutants has been already highlighted through this chapter. From the information summarized in Table 1 and Table 3 it is possible to infer that soil contamination with organic compounds, namely pesticides o polycyclic aromatic hydrocarbons, abates essential metabolic pathways such as the trycarboxylic acid cycle and the oxidative stress response, while lipid metabolism appears to be enhanced. However, the advance in the application of bioinformatics is providing further progress in terms of identification of specific biomarkers for risk assessment of individual target compounds. Thus, toxicity of endosulfan has been directly related with alterations of the GABA-glutamine cycle (Yuk et al., 2013), while chlorpyrifos depresses the Cori cycle and reduces the production of phospholipids, as indicated by lower levels of choline (Jones et al., 2012). Baylay et al. (2012) specifically relates chlorpyrifos toxicity to increased levels of fumarate, an intermediate of the trycarboxylic acid cycle. Research conducted with the same earthworm (*E. fetida*) and other families of organic compounds revealed a different metabolic response (Brown et al., 2010;Lankadurai et al., 2012), confirming the capability of metabolomics to discriminate the metabolic pathways involved in the response to a particular toxic compound. Moreover, the results strongly suggest that sets of biomarkers might be soon sufficiently reliable as for their implantation in in toxicity standardized test.

The relevance of these and future studies on the development of risk assessment strategies is aggravated by the inherent risk of soil contamination for human health. Soil contaminants may be responsible for health effects costing millions of euros. Health problems range from cancer (arsenic, asbestos, dioxins), to neurological damage and lower IQ (lead, arsenic), kidney disease (lead, mercury, cadmium), and skeletal and bone diseases (lead, fluoride, cadmium).

Overall, few studies have been conducted on the toxicity of complex chemical mixtures in soils. The effects of the soil and organisms within it upon organic pollutants are unknown. The data currently available correspond mostly to short-term studies and high level exposure of these chemicals, which is less relevant to the potential low-level, long term health impacts on living organisms near to contaminated soil.

7. METABOLITES RELATED TO SOIL CONTAMINATION WITH HEAVY METALS

The uptake of excess metal ions is toxic to most organisms, and the biochemical impact of metal ions on the cells varies with the chemistry of the element as their chemical nature. In plants, phytotoxicity of heavy metals in most parts can be attributed to symplastic accumulation of heavy metals, such as the cytosol and chloroplast stroma. Metal-induced changes in development are the result of either a direct and immediate impairment of metabolism or signaling processes that initiate adaptive or toxicity responses that need to be considered as active processes of the organism. Transport processes have been recognized as a central mechanism of metal detoxification and tolerance (Hall, 2002; Hall and Williams, 2003).

Some metals, for example, Zn and Cu, are essential for normal plant growth and development as they serve as structural and functional components of specific proteins. Other metals, for example, Cd and Pb, have no known function in plants although a Cd requirement for carbonic anhydrase from marine diatoms has been reported (Lane and Morel, 2000).

Upon exposure to metals, organisms often synthesize a set of diverse metabolites that accumulate to concentrations in the millimolar range, particularly specific amino acids, such as proline and histidine, peptides such as glutathione and phytochelatins (PC), and the amines spermine, spermidine, putrescine, nicotianamine, and mugineic acids that can be detected as response to these metals exposure. The advance of toxicogenomics in relation to organic contaminants is significantly ahead of the equivalent research in metal contaminated soil (Table 1). Nevertheless, research conducted up to date has yielded a number of biomarkers representative of the biological response of soil microorganisms to metals toxicity. Thus, soil contamination with Pb has been related with an enhancement of lipid metabolism (Sanchez-Hernandez, 2006) and more directly with reduction of tyrosine levels (Wu et al., 2013). Otherwise, Cd toxicity promotes the secretion of phytochelatins in *C. elegans*, likely at the expenses of the sulphur metabolism, as suggested by the reduction in cystathionine (Hughes et al., 2009), while the response of tomato plants to Cd involves several biochemical pathways (Hédiji et al., 2010). These examples illustrate the genuine specificity of biological reactions to different metals but also the variation in representative biomarkers among different organisms. Accordingly, exposure of *C. elegans*to Ni (Jones et al., 2012) yields a different metabolome than Cd since different biochemical pathways are affected.

In plants, data currently available demonstrate the significance of nitrogen-containing metabolites beyond phytochelatins and glutathione in plant response and acclimation to heavy metals. The various metal ions have specific chemical properties and induce distinct responses of adaptation and damage development. Thus, accumulating N-metabolites display a variety of functions, i.e. metal ion chelation, antioxidant defence, protection of macromolecules, and possibly signalling.

Proline is an extensively studied molecule in the context of plant responses to abiotic stresses. Up-regulation of proline is often encountered in plants under heavy metal stress, comparable to what occur under other abiotic stresses. When compared at equal toxic strength, proline accumulation decreased in the order Cd > Zn > Cu (Schat et al., 1997). In addition, it has been suggested different functions of proline under metal-stress, being involved in osmoregulation, metal chelation, antioxidant, and regulator of specific functions in plant morphogenesis.

Furthermore, Ni-hyperaccumulation has been specifically linked to histidine production (Krämer, 2005), particularly for *Saccharomyces cerevisiae* (Pearce and Sherman, 1999). The beneficial role of high histidine levels has been shown in transgenic *Arabidopsis thaliana* which accumulated about 2-fold higher histidine levels than wild- type plants and showed more than 10-fold increased biomass production in the presence of toxic Ni in the growth medium (Wycisk et al., 2004).. Moreover, cell surface-engineered yeast displaying a histidine oligopeptide (hexa-His) has been shown to adsorb 3–8 times more copper ions than the parent strain, being more resistant to Cu than the parent (Kuroda et al., 2002).

Otherwise, polyamine contents are altered in response to the exposure to heavy metals. Weinstein et al. (1986) showed an increment in putrescine content in Cd-treated oat seedlings and detached oat leaves with a marginal rise in spermidine and spermine content. They influence a variety of growth and development processes in plants and have been suggested to be a class of plant growth regulators and to act as second messengers (Kakkar and Sawhney, 2002). It has been suggested that they could stabilize and protect the membrane systems against the toxic effects of metal ions, particularly the redox active metals.

Overall, the number of studies remains rather scarce, and the preliminary results available in the literature merely constitute a launching platform for this promising research field.

8. FUTURE PERSPECTIVES

The main objective of metabolomics implementation in soil risk assessment is to meet the continuously increasing demand of safety data from human and ecological risk assessments. Accordingly, regulatory programs worldwide are currently incorporating tests with end-points that involve the effects of chemicals and the impact in specific metabolic pathways (Ankley et al., 2006). Toxicological end-points can be general biological responses such as survival or weight loss (Baylay et al., 2012), but specific biomarkers provide the accuracy that was classically elusive for test with living organisms

Several issues immediately arise from the summary here presented, such as the need to perform field toxicological test, with natural soils rather than use artificial soils, as was the case with some of the studies listed in Table 1. Ecotoxigenomics can also benefit from the incorporation of further analytical techniques. Techniques based on mass spectrometry are certainly required to understand the mechanisms involved in the alteration of metabolic pathways as response to toxicants. However, for screenings which merely require the detection of differences between metabolic phenotypes, optical methods such as FT-IR would be suitable, particularly if extremely high sample throughput is required (Bundy et al., 2009). Although no data was available in the existing literature, Figure 2 illustrates the change in the fingerprint of organic compounds in a soil amended with different sources of carbon collected 10 after the application. While some of the groups of compounds might be merely related to the sources of carbon added, the variations in the signal associated to polysaccharides (600-1000 cm^{-1}) can be associated to changes in the metabolic fingerprint of the soil system and therefore linked to microbiological activity in soil. Overall, the introduction of these results seeks to encourage further characterization of families of compounds in intact soil (or functional pools such as aggregates) in relation with soil processes, an approach that can find immediate application in the assessment of biological responses to toxic compounds in soil.

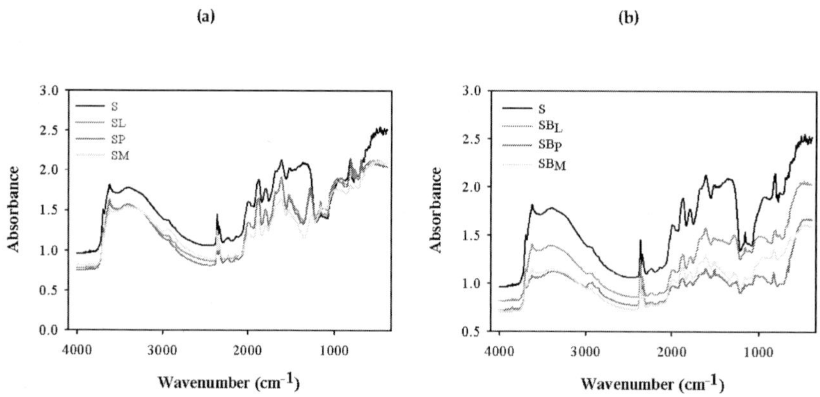

Figure 2. Absorption spectra obtained by Fourier transform infrared spectroscopy (FTIR) for an agricultural soil (S), soil amended with fresh residues (a): dry leaf litter (SL), peanut shell (SP), maize residue (SM), and soil amended with biochar (b) derived from those feedstocks (BL, BP or BM). Spectra presented (after 10 d incubation) are the average of 5 spectra obtained for different samples of each treatment. Hernandez-Soriano et al., unpublished data.

The variability of biological responses has been one of the main obstacles for their implementation in standardized risk assessment. However, the examination of changes in biological processes by accurate analytical techniques and powerful statistical tools has launched a new era in our understanding of the soil processes. The possibility of identifying the most sensitive metabolites for a certain toxicant and develop a tailored standardized test is the ultimate goal pursued.

ACKNOWLEDGEMENTS

MCH-S thanks The University of Queensland for a postdoctoral research fellowship. JCJ-L thanks the European research program Marie Curie (FP7-PEOPLE-2011-IOF) for his PIOF-GA-2011-301550 grant.

REFERENCES

1. Amiard-Triquet, C., Amiard, J.C., Rainbow, P.S., 2012. Ecological Biomarkers: Indicators of Ecotoxicological Effects. Taylor & Francis.

2. Ankley, G.T., Daston, G.P., Degitz, S.J., Denslow, N.D., Hoke, R.A., Kennedy, S.W., Miracle, A.L., Perkins, E.J., Snape, J., Tillitt, D.E., Tyler, C.R., Versteeg, D., 2006. Toxicogenomics in Regulatory Ecotoxicology. Environmental Science & Technology 40, 4055-4065.

3. Baylay, A.J., Spurgeon, D.J., Svendsen, C., Griffin, J.L., Swain, S.C., Sturzenbaum, S.R., Jones, O.A.H., 2012. A metabolomics based test of independent action and concentration addition using the earthworm Lumbricus rubellus. Ecotoxicology 21, 1436-1447.

4. Boer, M., Ellers, J., Gestel, C.M., Dunnen, J., Straalen, N., Roelofs, D., 2013. Transcriptional responses indicate attenuated oxidative stress in the springtail Folsomia candida exposed to mixtures of cadmium and phenanthrene. Ecotoxicology 22, 619-631.

5. Brown, S.A.E., McKelvie, J.R., Simpson, A.J., Simpson, M.J., 2010. 1H NMR metabolomics of earthworm exposure to sub-lethal concentrations of phenanthrene in soil. Environmental Pollution 158, 2117-2123.

6. Brown, S.A.E., Simpson, A.J., Simpson, M.J., 2009. 1H NMR metabolomics of earthworm responses to sub-lethal PAH exposure. Environmental Chemistry 6, 432-440.

7. Bundy, J., Davey, M., Viant, M., 2009. Environmental metabolomics: a critical review and future perspectives. Metabolomics 5, 3-21.

8. Bundy, J., Spurgeon, D., Svendsen, C., Hankard, P., Weeks, J., Osborn, D., Lindon, J., Nicholson, J., 2004. Environmental Metabonomics: Applying Combination Biomarker Analysis in Earthworms at a Metal Contaminated Site. Ecotoxicology 13, 797-806.

9. Bundy, J.G., Lenz, E.M., Bailey, N.J., Gavaghan, C.L., Svendsen, C., Spurgeon, D., Hankard, P.K., Osborn, D., Weeks, J.M., Trauger, S.A., Speir, P., Sanders, I., Lindon, J.C., Nicholson, J.K., Tang, H., 2002. Metabonomic assessment of toxicity of 4-fluoroaniline, 3,5-difluoroaniline and 2-fluoro-4-methylaniline to the earthworm Eisenia veneta (rosa): Identification of new endogenous biomarkers. Environmental Toxicology and Chemistry 21, 1966-1972.

10. Bundy, J.G., Osborn, D., Weeks, J.M., Lindon, J.C., Nicholson, J.K., 2001. An NMR-based metabonomic approach to the investigation of coelomic fluid biochemistry in earthworms under toxic stress. FEBS Letters 500, 31-35.

11. Dallinger-Marianne, J., 2000. Biomarkers in terrestrial invertebrates for ecotoxicological soil risk assessment. Reviews of Environmental Contamination and Toxicology 164, 93.

12. Drewes, C.D., Vining, E.P., 1984. In vivo neurotoxic effects of dieldrin on giant nerve fibers and escape reflex function in the earthworm, Eisenia foetida. Pesticide Biochemistry and Physiology 22, 93-103.

13. Edwards, C.A., Bater, J.E., 1992. The use of earthworms in environmental management. Soil Biology and Biochemistry 24, 1683-1689.

14. Ellis, J.K., Athersuch, T.J., Thomas, L.D.K., Teichert, F., Perez-Trujillo, M., Svendsen, C., Spurgeon, D.J., Singh, R., Jaerup, L., Bundy, J.G., Keun, H.C., 2012. Metabolic profiling detects early effects of environmental and lifestyle exposure to cadmium in a human population. Bmc Medicine 10.

15. Fitzpatrick, L.C., Sassani, R., Venables, B.J., Goven, A.J., 1992. Comparative toxicity of polychlorinated biphenyls to earthworms Eisenia foetida and Lumbricus terrestris. Environ Pollut 77, 65-69.

16. Gibb, J.O.T., Holmes, E., Nicholson, J.K., Weeks, J.M., 1997. Proton NMR spectroscopic studies on tissue extracts of invertebrate species with pollution indicator potential. Comparative Biochemistry and Physiology Part B: Biochemistry and Molecular Biology 118, 587-598.

17. Hall, J.L., 2002. Cellular mechanisms for heavy metal detoxification and tolerance. J Exp Bot 53, 1-11.

18. Hall, J.L., Williams, L.E., 2003. Transition metal transporters in plants. J Exp Bot 54, 2601-2613.

19. Hédiji, H., Djebali, W., Cabasson, C., Maucourt, M., Baldet, P., Bertrand, A., Boulila Zoghlami, L., Deborde, C., Moing, A., Brouquisse, R., Chaïbi, W., Gallusci, P., 2010. Effects of long-term cadmium exposure on growth and metabolomic profile of tomato plants. Ecotoxicology and Environmental Safety 73, 1965-1974.

20. Heimbach, F., 1988. A comparison of laboratory methods for toxicity testing with earthworms, In: Edwards, C.A., Neuhauser, E.F. (Eds.), Earthworms in waste and environmental management. SPB Academic Publishing, The Hague, The Netherlands, pp. 329-335.

21. Hughes, S.L., Bundy, J.G., Want, E.J., Kille, P., Stürzenbaum, S.R., 2009. The Metabolomic Responses of Caenorhabditis elegans to Cadmium Are Largely Independent of Metallothionein Status, but Dominated by Changes in Cystathionine and Phytochelatins. Journal of Proteome Research 8, 3512-3519.

22. Jimenez-Lopez, J., Kotchoni, S., Hernandez-Soriano, M., Gachomo, E., Alché, J., 2013. Structural functionality, catalytic mechanism modeling and molecular allergenicity of phenylcoumaran benzylic ether reductase, an olive pollen (Ole e 12) allergen. Journal of Computer-Aided Molecular Design 27, 873-895.

23. Jones, O.A.H., Sdepanian, S., Lofts, S., Svendsen, C., Spurgeon, D.J., Maguire, M.L., Griffin, J.L., 2014. Metabolomic analysis of soil communities can be used for pollution assessment. Environmental Toxicology and Chemistry 33, 61-64.

24. Jones, O.A.H., Spurgeon, D.J., Svendsen, C., Griffin, J.L., 2008. A metabolomics based approach to assessing the toxicity of the polyaromatic hydrocarbon pyrene to the earthworm Lumbricus rubellus. Chemosphere 71, 601-609.

25. Jones, O.A.H., Swain, S.C., Svendsen, C., Griffin, J.L., Sturzenbaum, S.R., Spurgeon, D.J., 2012. Potential New Method of Mixture Effects Testing Using Metabolomics and Caenorhabditis elegans. Journal of Proteome Research 11, 1446-1453.

26. Kakkar, R.K., Sawhney, V.K., 2002. Polyamine research in plants – a changing perspective. Physiologia Plantarum 116, 281-292.

27. Kammenga, J.E., Dallinger, R., Donker, M.H., Kohler, H.R., Simonsen, V., Triebskorn, R., Weeks, J.M., 2000. Biomarkers in terrestrial invertebrates for ecotoxicological soil risk assessment. Rev Environ Contam Toxicol 164, 93-147.

28. Krämer, U., 2005. Phytoremediation: novel approaches to cleaning up polluted soils. Current Opinion in Biotechnology 16, 133-141.

29. Kuroda, K., Ueda, M., Shibasaki, S., Tanaka, A., 2002. Cell surface-engineered yeast with ability to bind, and self-aggregate in response to, copper ion. Applied Microbiology and Biotechnology 59, 259-264.

30. Kwon, Y.-K., Jung, Y.-S., Park, J.-C., Seo, J., Choi, M.-S., Hwang, G.-S., 2012. Characterizing the effect of heavy metal contamination on marine mussels using metabolomics. Marine Pollution Bulletin 64, 1874-1879.

31. Lane, T.W., Morel, F.M.M., 2000. A biological function for cadmium in marine diatoms. Proceedings of the National Academy of Sciences 97, 4627-4631.

32. Lankadurai, B.P., Simpson, A.J., Simpson, M.J., 2012. 1H NMR metabolomics of Eisenia fetida responses after sub-lethal exposure to perfluorooctanoic acid and perfluorooctane sulfonate. Environmental Chemistry 9, 502-511.

33. Lankadurai, B.P., Wolfe, D.M., Simpson, A.J., Simpson, M.J., 2011. 1H NMR-based metabolomics of time-dependent responses of Eisenia fetida to sub-lethal phenanthrene exposure. Environmental Pollution 159, 2845-2851.

34. Lenz, E.M., Weeks, J.M., Lindon, J.C., Osborn, D., Nicholson, J.K., 2005. Qualitative high field 1H-NMR spectroscopy for the characterization of endogenous metabolites in earthworms with biochemical biomarker potential. Metabolomics 1, 123-136.

35. McKelvie, J.R., Wolfe, D.M., Celejewski, M.A., Alaee, M., Simpson, A.J., Simpson, M.J., 2011. Metabolic responses of Eisenia fetida after sub-lethal exposure to organic contaminants with different toxic modes of action. Environmental Pollution 159, 3620-3626.

36. McKelvie, J.R., Yuk, J., Xu, Y., Simpson, A.J., Simpson, M.J., 2009. H-1 NMR and GC/MS metabolomics of earthworm responses to sub-lethal DDT and endosulfan exposure. Metabolomics 5, 84-94.

37. Metwally, A., Safronova, V.I., Belimov, A.A., Dietz, K.J., 2005. Genotypic variation of the response to cadmium toxicity in Pisum sativum L. J Exp Bot 56, 167-178.

38. Miller, M.G., 2007. Environmental Metabolomics: A SWOT Analysis (Strengths, Weaknesses, Opportunities, and Threats). Journal of Proteome Research 6, 540-545.

39. Morrison, N., Bearden, D., Bundy, J., Collette, T., Currie, F., Davey, M., Haigh, N., Hancock, D., Jones, O.H., Rochfort, S., Sansone, S.-A., Štys, D., Teng, Q., Field, D., Viant, M., 2007. Standard reporting requirements for biological samples in metabolomics experiments: environmental context. Metabolomics 3, 203-210.

40. OECD, 1984. Earthworm, acute toxicity tests. Guideline for testing chemicals. No. 207. OECD, Paris, France.

41. OECD, 2004. Earthworm reproduction test. Guideline for testing chemicals. No. 222. OECD, Paris, France.

42. Paschalidis, K.A., Roubelakis-Angelakis, K.A., 2005. Sites and regulation of polyamine catabolism in the tobacco plant. Correlations with cell division/expansion, cell cycle progression, and vascular development. Plant Physiol 138, 2174-2184.

43. Pearce, D.A., Sherman, F., 1999. Toxicity of copper, cobalt, and nickel salts is dependent on histidine metabolism in the yeast Saccharomyces cerevisiae. J Bacteriol 181, 4774-4779.

44. Rauser, W., 1995. Phytochelatins and related peptides. Structure, biosynthesis, and function. Plant Physiology 109, 1141-1149.

45. Regoli, F., Gorbi, S., Frenzilli, G., Nigro, M., Corsi, I., Focardi, S., Winston, G.W., 2002. Oxidative stress in ecotoxicology: from the analysis of individual antioxidants to a more integrated approach. Marine Environmental Research 54, 419-423.

46. Rochfort, S.J., Ezernieks, V., Yen, A.L., 2009. NMR-based metabolomics using earthworms as potential indicators for soil health. Metabolomics 5, 95-107.

47. Sanchez-Hernandez, J.C., 2006. Earthworm Biomarkers in Ecological Risk Assessment, In: Ware, G., Whitacre, D., Albert, L., Voogt, P., Gerba, C., Hutzinger, O., Knaak, J., Mayer, F., Morgan, D.P., Park, D., Tjeerdema, R.,

Yang, R.H., Gunther, F. (Eds.), Reviews of Environmental Contamination and Toxicology. Springer New York, pp. 85-126.

48. Sanchez-Hernandez, J.C., 2010. Environmental applications of earthworm esterases in the agroecosystem. Journal of Pesticide Science 35, 290-301.

49. Schat, H., Sharma, S., Vooijs, R., 1997. Heavy metal-induced accumulation of free proline in metal-tolerant and a nontolerant ecotype of Silene vulgaris. Physiologia Plantarum 101, 477-482.

50. Schutzendubel, A., Polle, A., 2002. Plant responses to abiotic stresses: heavy metal-induced oxidative stress and protection by mycorrhization. J Exp Bot 53, 1351-1365.

51. Shulaev, V., 2006. Metabolomics technology and bioinformatics. Briefings in Bioinformatics 7, 128-139.

52. Singh, O.V., 2006. Proteomics and metabolomics: The molecular make-up of toxic aromatic pollutant bioremediation. PROTEOMICS 6, 5481-5492.

53. Siripornadulsil, S., Traina, S., Verma, D.P., Sayre, R.T., 2002. Molecular mechanisms of proline-mediated tolerance to toxic heavy metals in transgenic microalgae. Plant Cell 14, 2837-2847.

54. Smilde, A.K., Westerhuis, J.A., Hoefsloot, H.C.J., Bijlsma, S., Rubingh, C.M., Vis, D.J., Jellema, R.H., Pijl, H., Roelfsema, F., Greef, J., 2010. Dynamic metabolomic data analysis: a tutorial review. Metabolomics 6, 3-17.

55. Smirnoff, N., 1993. The role of active oxygen in the response of plants to water deficit and desiccation. New Phytologist 125, 27-58.

56. Smith, C.A., O'Maille, G., Want, E.J., Qin, C., Trauger, S.A., Brandon, T.R., Custodio, D.E., Abagyan, R., Siuzdak, G., 2005. METLIN: a metabolite mass spectral database. Ther Drug Monit 27, 747-751.

57. Snape, J.R., Maund, S.J., Pickford, D.B., Hutchinson, T.H., 2004. Ecotoxicogenomics: the challenge of integrating genomics into aquatic and terrestrial ecotoxicology. Aquatic Toxicology 67, 143-154.

58. Spurgeon, D.J., Morgan, A.J., Kille, P., 2008. Current research in soil invertebrate ecotoxicogenomics, In: Hogstrand, C., Kille, P. (Eds.), Comparative Toxicogenomics, pp. 133-163.

59. Tremaroli, V., Workentine, M.L., Weljie, A.M., Vogel, H.J., Ceri, H., Viti, C., Tatti, E., Zhang, P., Hynes, A.P., Turner, R.J., Zannoni, D., 2009.

Metabolomic investigation of the bacterial response to a metal challenge. Appl Environ Microbiol 75, 719-728.

60. Trygg, J., Holmes, E., Lundstedt, T., 2006. Chemometrics in Metabonomics. Journal of Proteome Research 6, 469-479.

61. Van Gestel, C.A.M., Dirven-Van Breemen, E.M., Baerselman, R., Emans, H.J.B., Janssen, J.A.M., Postuma, R., Van Vliet, P.J.M., 1992. Comparison of sublethal and lethal criteria for nine different chemicals in standardized toxicity tests using the earthworm Eisenia andrei. Ecotoxicology and Environmental Safety 23, 206-220.

62. van Gestel, C.A.M., van Dis, W.A., van Breemen, E.M., Sparenburg, P.M., 1989. Development of a standardized reproduction toxicity test with the earthworm species Eisenia fetida andrei using copper, pentachlorophenol, and 2,4-dichloroaniline. Ecotoxicology and Environmental Safety 18, 305-312.

63. van Ravenzwaay, B., Cunha, G.C.-P., Leibold, E., Looser, R., Mellert, W., Prokoudine, A., Walk, T., Wiemer, J., 2007. The use of metabolomics for the discovery of new biomarkers of effect. Toxicology Letters 172, 21-28.

64. van Ravenzwaay, B., Herold, M., Kamp, H., Kapp, M.D., Fabian, E., Looser, R., Krennrich, G., Mellert, W., Prokoudine, A., Strauss, V., Walk, T., Wiemer, J., 2012. Metabolomics: A tool for early detection of toxicological effects and an opportunity for biology based grouping of chemicals—From QSAR to QBAR. Mutation Research/Genetic Toxicology and Environmental Mutagenesis 746, 144-150.

65. Vega-López, A., Ayala-López, G., Posadas-Espadas, B.P., Olivares-Rubio, H.F., Dzul-Caamal, R., 2013. Relations of oxidative stress in freshwater phytoplankton with heavy metals and polycyclic aromatic hydrocarbons. Comparative Biochemistry and Physiology Part A: Molecular & Integrative Physiology 165, 498-507.

66. Viant, M.R., 2009. Applications of metabolomics to the environmental sciences. Metabolomics 5, 1-2.

67. Viant, M.R., Rosenblum, E.S., Tjeerdema, R.S., 2003. NMR-Based Metabolomics: A Powerful Approach for Characterizing the Effects of Environmental Stressors on Organism Health. Environmental Science & Technology 37, 4982-4989.

68. Warne, M., Lenz, E.M., Osborn, D., Weeks, J., Nicholson, J., 2000. An NMR-based metabonomic investigation of the toxic effects of 3-trifluoromethyl-aniline on the earthworm Eisenia veneta. Biomarkers 5, 56-72.

69. Weinstein, L.H., Kaur-Sawhney, R., Rajam, M.V., Wettlaufer, S.H., Galston, A.W., 1986. Cadmium-induced accumulation of putrescine in oat and bean leaves. Plant Physiol 82, 641-645.

70. Whitfield Åslund, M., Simpson, A., Simpson, M., 2011. 1H NMR metabolomics of earthworm responses to polychlorinated biphenyl (PCB) exposure in soil. Ecotoxicology 20, 836-846.

71. Whitfield Aslund, M.L., McShane, H., Simpson, M.J., Simpson, A.J., Whalen, J.K., Hendershot, W.H., Sunahara, G.I., 2012. Earthworm sublethal responses to titanium dioxide nanomaterial in soil detected by (1)H NMR metabolomics. Environ Sci Technol 46, 1111-1118.

72. Wu, H., Liu, X., Zhao, J., Yu, J., 2013. Regulation of Metabolites, Gene Expression, and Antioxidant Enzymes to Environmentally Relevant Lead and Zinc in the Halophyte Suaeda salsa. Journal of Plant Growth Regulation 32, 353-361.

73. Wycisk, K., Kim, E.J., Schroeder, J.I., Krämer, U., 2004. Enhancing the first enzymatic step in the histidine biosynthesis pathway increases the free histidine pool and nickel tolerance in Arabidopsis thaliana. FEBS Letters 578, 128-134.

74. Yu, Y.L., Wu, X.M., Li, S.N., Fang, H., Zhan, H.Y., Yu, J.Q., 2006. An exploration of the relationship between adsorption and bioavailability of pesticides in soil to earthworm. Environmental Pollution 141, 428-433.

75. Yuk, J., McKelvie, J.R., Simpson, M.J., Spraul, M., Simpson, A.J., 2010. Comparison of 1-D and 2-D NMR techniques for screening earthworm responses to sub-lethal endosulfan exposure. Environmental Chemistry 7, 524-536.

76. Yuk, J., Simpson, M.J., Simpson, A.J., 2013. 1-D and 2-D NMR-based metabolomics of earthworms exposed to endosulfan and endosulfan sulfate in soil. Environmental Pollution 175, 35-44.

77. Zhang, Q., Bhattacharya, S., Andersen, M.E., Conolly, R.B., 2010. Computational Systems Biology and Dose-Response Modeling in Relation to New Directions in Toxicity Testing. Journal of Toxicology and Environmental Health, Part B 13, 253-276.

Chapter 3

Influence of Oil Contamination on Physical and Biological Properties of Forest Soil After Chainsaw Use

Anna Klamerus Iwan , Ewa Błońska, Jarosław Lasota, Agnieszka Kalandyk, Piotr Waligórski

Faculty of Forestry, Department of Forest Engineering, University of Agriculture in Kraków

ABSTRACT

Forestry works using chainsaws result in up to 7 million liters of various mineral oils being soaked annually into forest soils. These substances, containing a complex mixture of polycyclic aromatic hydrocarbons (PAHs), are highly toxic. The aim of the study was to determine the effect of oil contamination with PAHs on the physical and biological properties of forest soils. The study area was located in southern Poland in the Miechów forest district. The experiment was conducted on four treatment blocks with various amounts of oil addition. The study included the determination of PAH content, dehydrogenase and urease activity, and biomass of earthworms. Physical properties were determined using the dryer method and Kopecky rings of 250 cm^3 volume. The results obtained confirmed the hypothesis that oil contamination with PAHs modified the physical properties of forest soils and oil had a negative impact on enzyme activity in soil. Enzyme activity in the studied soils was negatively correlated with PAH content. Earthworm population density reflected the contamination level of oil-contaminated soils.

KEYWORDS

Polycyclic aromatic hydrocarbons (PAHs), Enzyme activity, Earthworms, Mechanization in forestry, Principal component analysis (PCA)

1 INTRODUCTION

Soil contamination by oil from chainsaws is a serious problem in forest management. Most jobs performed in forests require the use of chainsaws. It is estimated that with logging in Poland yielding around 30 million $m^3$of wood per year, approximately 6 million dm^3 of oil—both original, first-time use oil, and re-used oil after reprocessing—penetrates into the environment. Environmental contamination by oil is especially dangerous in the case of used oil because of its strong toxic properties (Giefing 1991; Rudko and Rybczyński 2010). Mechanization in forestry leads to the emission of pollutants into the natural environment, including polycyclic aromatic hydrocarbons (PAHs). Oils are the source of PAHs and the PAHs contained in oil penetrate into the soil environment (Lipińska et al. 2013). The soil system is considered the most important long-term repository for PAHs, and it is also considered to be a steady indicator of the state of environmental pollution (Nam et al. 2003; Mueller and Shann 2006). PAHs affect the activity of soil enzymes, which can be used to evaluate soil microbial properties (Shen et al. 2006). The activity of soil enzymes is one of the approved parameters used for the evaluation of soil quality polluted with organic compounds, including PAHs (Lipińska et al. 2014). The activity of urease appears to be more sensitive to pollution than other soil enzymes (Bååth 1989). Dehydrogenase activity is the most frequently used test for determining the influence of various pollutants on the microbiological quality of soil (Sannino and Gianfreda 2001; Baran et al. 2004).

Contamination with hydrocarbons can have a profound effect on soil fauna (Dendooven et al. 2011). Several authors reported the negative effect of PAHs on the survival and reproduction of earthworms (Brown et al.2004; Contreras-Ramos et al. 2006).

Oil pollution might affect soil physical properties. Pore spaces might be clogged, which could reduce soil aeration and water infiltration and increase bulk density, subsequently affecting plant growth. Oils that are denser than water might reduce and restrict soil permeability (Abosede 2013).

Beech (*Fagus sylvatica*) is a forest species of high economic importance. In recent years, this species has frequently been introduced into Polish forests. After harvesting of trees, foresters plant new seedlings and they want to know the short-term negative effects of the mineral oil from chainsaw use on soil properties. The growth and quality of seedlings depends on the soil properties.

The aim of the study was to determine the effect of oil contamination with PAHs on the physical and biological properties of forest soils characterized by the same pH, clay, and carbon contents. The study involved the evaluation of the influence of different amounts of oil on soil moisture and aeration conditions, enzyme activity, and density of earthworms. We hypothesized that: (1) contamination of soil with PAHs would have a negative impact on dehydrogenase and urease activity of forest soil; (2) earthworm population density would reflect the state of oil-contaminated soils; (3) contamination by oil containing PAHs would modify the moisture and aeration characteristics of forest soils.

2 MATERIALS AND METHODS

2.1 Soil Sampling Sites

Sample plots were located in southern Poland in the vicinity of Kraków, within the Miechów forest district (50.3000°N, 19.9833°E). The study area is characterized by the following climate conditions: the average precipitation in the months of July and August is 87.5 mm and the average temperature is 16.6 °C. Sample plots were located in an area with a predominance of highland loess of aeolian origin. The test area was dominated by Cambisols (WRB 2002). Table 1 shows the soil properties for the test site. A detailed description of the soil profile was carried out; samples were taken from each genetic horizon to characterize the basic soil properties (soil pH in H_2O and 1 M KCl solutions, hydrolytic acidity, exchangeable acidity, exchangeable Al, total nitrogen (N) and carbon (C) contents and calculated C/N ratio, granulometric composition).

Beech (*F. sylvatica*) was the dominant species in the study area stands. The study area was divided into four blocks (I, II, III, and IV) of 2 × 2 m (4 m^2). Oil (50 g/m^2 (D1)) was administered to the first block, 100 g/m^2 (D2) to the second block, and 200 g/m^2 (D3) to the third block. Block IV served as a control (*C*). The experiment was carried out during the vegetation period, in the months of July and August 2014. Various amounts of oil were added to the soil in July, and samples for laboratory analysis were collected a month later. Oil was applied to the soil using a sprinkler.

Table 1. Properties of investigated soil

Table 1 Properties of investigated soil

Depth	*C*	*N*	C/N	pH H_2O	pH KCl	*Y*	H_w	Al	*S*	*V*	Sand	Dust	Clay	Humus type	TSL
0–15	4.81	0.54	9	4.82	4.12	8.12	0.83	0.41	6.7	35	5	88	7	Mull	Lwyżśw
15–45	0.54	0.02	27	4.61	3.74	4.45	4.31	4.22	1.1	18	7	86	7		
45–100	n.d.	n.d.	n.d.	4.91	3.74	4.89	4.21	4.18	7.5	58	7	83	10		
100–120	n.d.	n.d.	n.d.	5.23	3.98	2.21	0.84	0.63	8.8	80	7	84	9		

C total organic carbon (%), N total nitrogen (%), C/N C/N ratio, S sum of exchangeable base cations (cmol(+)kg^{-1}), V saturation of base cations (%), Y hydrolytic acidity (cmol(+)kg^{-1}), Hw exchangeable acidity (cmol(+)kg^{-1}), Al exchangeable aluminum (mg 100 g^{-1})

TSL type of forest site, Lwyżśw upland broadleaf forest site, *n.d.* no determined

2.2 Physical Analyses

Soil moisture (Wv expressed as % of volume, or Ww expressed as % of weight), capillary water capacity (CWCw expressed as % of weight, or CWCv expressed as % of volume), and bulk density (Bd) were determined using the dryer method with Kopecky rings of 250 cm^3 volume (Ostrowska et al. 1991). Density of soil solid phase was determined by pycnometry, while total porosity was calculated on the basis of density of the solid phase and soil density. Air-filled porosity was calculated based on the total porosity and capillary water capacity.

2.3 Biological Analyses

Dehydrogenase (EC 1.1.1.1) and urease (EC 3.5.1.5) activities were determined. Five replicate soil samples were taken from each block at a depth of 0–10 cm for a total of 20 samples. Dehydrogenase activity was detected using Lenhard's method according to the Casida procedure (Alef and Nannipieri 1995); enzyme activity was expressed as mg triphenyl formazan (TPF) formed per 100 g soil per 24 h. Urease activity was determined using the method of Tabatabai and Bremner (1972; as cited in Alef and Nannipieri 1995); enzyme activity was expressed as μg $N\text{-}NH_4$ formed per 1 g of soil per 2 h.

To determine earthworm density, two methods were used: manual sorting and the formalin extraction method (Clapperton et al. 2007). Five replicates were used for the calculation of earthworm density and biomass for each treatment.

2.4 Determination of PAHs

The soil was air dried at 40 °C for 48 h and then ground in a mortar. Approximately 5 g was weighed from each sample, and then PAHs were extracted using 35 mL of *n*-hexane. The samples were sonicated and centrifuged (4000*g*, 5 min), the supernatant was collected, and the extraction was repeated with the application of 20 mL of *n*-hexane. Both supernatants were combined and evaporated to dryness using a rotary evaporator. The residue was dissolved in acetonitrile and analyzed for PAHs by high-performance liquid chromatography (HPLC). The Agilent Technologies 1260 system was equipped with a fluorescence detector and Agilent Technologies Eclipse XDB-C18 5 μm, 4.6 × 150 mm HPLC column. The mobile phases were water (A) and acetonitrile (B) at a flow rate of 1 mL/min. The sample injection volume was 5 μL. Compounds were eluted using the following gradient: 0–5 min 20:80 A:B→13:87 A:B, 5–7.5 min held at 13:87 A:B, 7.5–15 min 13:87 A:B→0:100 A:B. Ten-point calibration curves were prepared for the following compounds (retention time and excitation and emission wavelengths are provided in brackets): phenanthrene (4.77 min, 252 nm, 365 nm), pyrene (6 min, 240 nm, 390 nm), chrysene (7 min, 270 nm, 380 nm), benzo[a]pyrene (10 min, 265 nm, 410 nm) and dibenz[a,h]anthracene (11 min, 295 nm, 405 nm). These compounds were chosen because of their high content in raw oil and petrochemical products. We also analyzed the PAHs in oil. One hundred microliters of oil was extracted with 10 mL acetonitrile, vortexed and sonicated for 20 min. The sample was then centrifuged (20,000*g* for 20 min), and the supernatant was analyzed with the same HPLC method as described above.

2.5 Statistical Analyses

To reduce the number of variables in the statistical data set and to visualize the multivariate data set as a set of coordinates in a high-dimensional data space, principal component analysis (PCA) was used. The PCA method was also used to interpret other factors, depending on the type of data set. In PCA analysis, the physical properties and enzyme activities of soil were used. Differences between the mean values in soil groups were evaluated with the Tukey's test ($P < 0.05$). All statistical analyses were performed with Statistica 10 software (2010).

3 RESULTS

3.1 Soil Physical Properties

Changes in selected properties of soil resulting from the addition of chainsaw oil to the soil were observed. Slightly lower moisture content was noted in the soil samples collected from the block where the highest amount of oil was added. However, the relative moisture expressed in relation to the dry sample weight was not significantly different between treatments (Table 2). Capillary water content increased slightly in correlation with the amount of oil added to the soil (Table 2). Soil bulk density increased with increased soil contamination with chainsaw oil and total porosity and air-filled porosity simultaneously decreased. Oil at 100 g/m^2 caused a decrease of 4 % in air-filled porosity, and the highest amount of oil (200 g/m^2) resulted in a decrease in air-filled porosity of nearly 10 % compared with unpolluted soil (Table 2).

Table 2. Physical properties of soil with different dose of oil

Dose of oil	Wv	Ww	Pwk	Pwv	Bd	Pt	Pa
D1	$10.42^{a}\pm1.39$	$7.0^{a}\pm1.03$	$18.8^{a}\pm1.91$	$27.8^{a}\pm2.46$	$1.48^{a}\pm0.02$	$39.5^{b}\pm0.91$	$11.7^{bc}\pm1.67$
D2	$10.94^{a}\pm0.73$	$7.0^{a}\pm0.62$	$18.4^{a}\pm1.12$	$28.6^{a}\pm1.22$	$1.51^{ab}\pm0.09$	$38.4^{ab}\pm3.60$	$9.7^{b}\pm2.86$
D3	$9.74^{a}\pm0.97$	$6.5^{a}\pm0.76$	$19.7^{a}\pm1.97$	$29.7^{a}\pm1.50$	$1.59^{b}\pm0.04$	$35.2^{a}\pm1.70$	$5.5^{a}\pm1.89$
C	$11.26^{a}\pm2.47$	$6.9^{a}\pm1.22$	$17.3^{a}\pm1.26$	$27.9^{a}\pm1.53$	$1.42^{a}\pm0.08$	$41.9^{b}\pm0.80$	$14.0^{c}\pm1.41$

Different small letters in the upper index of the mean values mean significant differences

Dose of oil: D1 50 g/m^2, D2 100 g/m^2, D3 200 g/m^2, C 0 (control); Wv relative moisture content in volume percent, Ww relative moisture content in weight percent, Pwk capillary water capacity in weight percent, Pwv capillary water capacity in volume percent, Bd bulk density of soil (g cm^{-3}), Pt total porosity of soil in volume percent, Pa non-capillary porosity in volume percent

Water resistance of soil aggregates is another property of soil that can be modified by the presence of oily substances containing aromatic hydrocarbons. The addition of oil to soil resulted in the increased water resistance of different size aggregates; this influence was stronger the higher the amount of oil introduced into the soil (Table 3). Oil at 200 g/m^2 resulted in almost complete resistance of aggregates to disintegration in water, even after 24 h immersion in water. In contrast, small and medium-sized aggregates from soil not treated with oil disintegrated by 40–60 % after 1 h immersion and after 24 h immersion, no permanent aggregates were noted (Table 3).

Table 3. Water resistant of aggregate in soil with different of dose oil

Measurement time (h)	Dose of oil	Diameter of soil aggregates (mm)			
		>10	5–10	3–5	2–3
1	D1	75	80	80	80
	D2	75	60	60	80
	D3	100	100	100	100
	C	75	60	40	60
5	D1	75	80	60	60
	D2	75	40	40	60
	D3	100	100	100	100
	C	75	40	20	40
24	D1	75	60	0	40
	D2	75	40	20	0
	D3	100	100	75	100
	C	50	20	0	0

Dose of oil: D1 50 g/m^2, D2 100 g/m^2, D3 200 g/m^2, C 0 (control)

3.2 Soil Biological Properties

Oil contamination of soil had significant effects on soil biological activity. A strong reduction in dehydrogenase activity was observed at 100 g/m^2, and at 200 g/m^2, dehydrogenase activity was reduced by approximately 50 % compared with the activity in the control soil (Table 4). A similar pattern was observed for urease activity. The lowest amount of oil (50 g/m^2) resulted in a slight decrease in urease activity, while activity was reduced by 40 and 50 % at 100 and 200 g/m^2, respectively, compared with the control. The chainsaw oil used in the experiment had a major influence on the density and biomass of earthworms. Even the smallest amount of oil (50 g/m^2) caused a decline in the population of earthworms from >40 specimens/m^2 in the control block to ~9 specimens/m^2. Oil at 100 g/m^2 affected the species composition of earthworms; this was confirmed by lower biomass of these animals (with similar numbers, the biomass decreased from >3.0 to 0.4 g/m^2; Table 4).

Table 4. Enzyme activity and quantity of earthworms in the soil with different dose of oil

Dose of oil	DH	UR	DE	BE
D1	25.96^c ±1.50	0.20^c ±0.02	8.67^c ±0.58	3.37^b ±0.00
D2	19.11^b ±1.70	0.15^b ±0.01	7.33^b ±0.58	0.41^a ±0.06
D3	15.93^a ±3.01	0.12^a ±0.01	0.00^a ±0.00	0.00^a ±0.00
C	27.80^c ±4.68	0.25^d ±0.00	43.33^d ±0.58	12.08^c ±0.86

Different small letters in the upper index of the mean values mean significant differences

Dose of oil: D1 50 g m^{-2}, D2 100 g m^{-2}, D3 200 g m^{-2}, C 0 (control), *DH* dehydrogenase activity (μM TPF kg^{-1} soil h^{-1}), *UR* urease activity (mM N-NH$_4$kg^{-1} soil h^{-1}), *DE* density earthworms (the number of pieces m^{-2}), *BE* biomass of earthworms (g m^{-2})

3.3 Soil PAH Content

The contents of selected PAHs in oil and soil on the first day of the experiment were presented in Table 5. The PAH contents in oil-contaminated soils after 1 month were presented in Table 6. Among the tested soils, the highest hydrocarbon contents were recorded for chrysene and pyrene. Hydrocarbon contents were higher in oil-contaminated soils than in the control soil. The highest concentrations of chrysene (0.1756 μg/g), phenanthrene, pyrene, and benzo(a)pyrene were observed in soils with highest amount of oil (D3). The concentration of phenanthrene in soil after 1 month was the most reduced compared with the starting concentration, indicating substantial degradation of this compound during the incubation period. Pyrene, chrysene, and benzo(a)pyrene were strongly absorbed to soil and thus poorly bioavailable for degradation (Tables 5 and 6).

Table 5. Selected polycyclic aromatic hydrocarbons in oil (µg ml) and soil with different dose of oil in first day experiment (µg g of soil)

	Phenanthrene	Pyrene	Chryzene	Benzo-a-pyrene	Dibenzo-a,h-anthracene
D1	0.0547	0.0413	0.1129	0.0042	0.000
D2	0.0785	0.0527	0.1326	0.0053	0.000
D3	0.1262	0.0756	0.1721	0.0075	0.000
PAHs in oil	6.1000	2.9301	5.0500	0.2800	0.000

Dose of oil: D1 50 g/m^2, D2 100 g/m^2, D3 200 g/m^2, C 0 (control)

Table 6. Selected polycyclic aromatic hydrocarbons in soil with different dose of oil after 1 month

Dose of oil	Phenanthrene	Pyrene	Chryzene	Benzo-a-pyrene	Dibenzo-a,h-anthracene
D1	0.0356	0.0393	0.1157	0.0040	0.0088
D2	0.0337	0.0500	0.1275	0.0020	0.0082
D3	0.0414	0.0622	0.1756	0.0062	0.0156
C	0.0308	0.0298	0.0931	0.0031	0.0056

Dose of oil: D1 50 g/m^2, D2 100 g/m^2, D3 200 g/m^2, C 0 (control)

3.4 PCA Analysis

Two main factors were selected that explained 82.11 % of the total variance of the variables (Fig. 1). Factor 1, which represented 71.85 % of the variation, was strongly correlated with enzyme activity, biomass and density of earthworms, and PAH content. Factor 2, which explained 10.26 % of the variation, was correlated with the soil water content and capillary water capacity.

Biological properties of the studied soils were negatively correlated with PAH content. Enzyme activity and biomass and density of earthworms were positively correlated with porosity (Fig. 1).

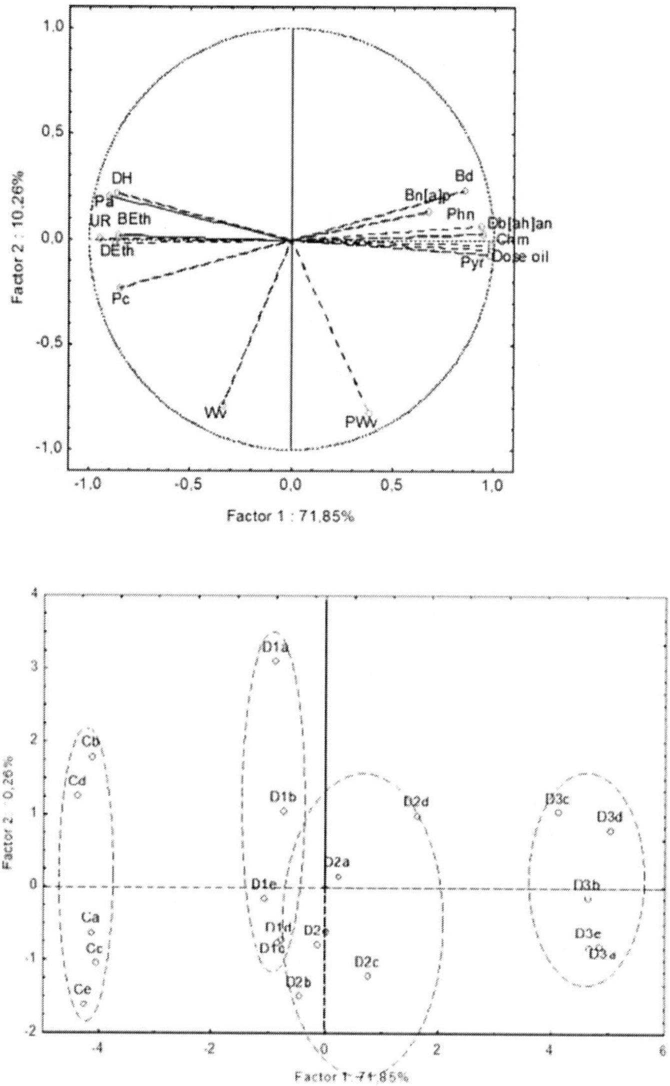

Figure. 1 Factorial plan and projection of variables in the soil properties on the factor-plane 1 × 2

4 DISCUSSION

During this study, changes in soil physical properties because of the application of oil were observed. Slightly lower moisture content (expressed as soil volume %) was recorded in the soil samples collected from the block where the highest amount of nebulized oil was applied. Capillary water capacity slightly increased with the amount of oil introduced into the soil. It is difficult to say whether this trend was directly related to the change in physical properties of soil when oil was introduced into it, or whether the differences were within the acceptable range of variation for a single stand of trees under natural conditions. The increase in capillary water capacity correlated with the increasing amount of oil added to the soil could be explained by a decrease in the diameter of larger capillaries, and consequently stronger water-holding capacity of these capillaries, or inhibition of their drainage capacity. Khamehchiyan et al. (2007) and Kuyukina et al. (2005) reported that high concentrations of crude oil in soil could clog soil pores and reduce water and oxygen penetration. According to Abosede (2013), crude oil might have negative effects on some soil physical properties including decreased pore spaces, saturated hydraulic conductivity, and increased bulk density. In this study, soil contamination with chainsaw mineral oil increased the soil bulk density, with simultaneous deterioration of total porosity and air-filled porosity. Oil contamination can lead to anaerobic conditions in soil by smothering soil particles and blocking air diffusion into soil pores, with subsequent effects on soil microbial communities (Townsend et al. 2003; Labud et al. 2007; Sutton et al. 2013). Water resistance of soil aggregates is another property of soil that can be modified by the presence of oily substances containing aromatic hydrocarbons. In addition, crude oil-contaminated soils are generally more hydrophobic than pristine sites (Quyum et al. 2002; Aislabie et al. 2004).

The negative impact of soil contamination by petroleum products and PAHs on soil microorganisms and their metabolic activity has been confirmed by numerous experiments carried out under controlled conditions. The impact of individual PAHs and PAH mixtures on the microbiological condition of agricultural soils has been extensively investigated (Eschenbach et al. 1991; Maliszewska-Kordybach et al. 2000; Kucharski et al.2000, 2004; Klimkowicz-Pawlas and Maliszewska-Kordybach 2003). A very important factor affecting the intensity of the influence of PAHs on soil microorganisms is the quantity and quality of soil organic matter. The bioavailability and environmental persistence of PAHs are most affected by soil organic matter content. Soil organic matter content also determines the enzyme activity and abundance of microorganisms (Veres et al. 2015; Kotroczó et al. 2014). Increased organic matter content in soil promotes stronger adsorption of PAHs, which reduces their content in the aqueous phase of the soil, and thus their bioavailability (Maliszewska-Kordybach et al. 2000). The bioavailable fraction of PAHs in soils polluted over a long period of time represents only a small percentage of their total content because of sorption by soil humic compounds (Swartz et al. 1995).

Dehydrogenases are intracellular enzymes that undergo rapid degradation after cell death, and are therefore not accumulated in the soil. Because of this property, their activity is often used in ecotoxicological research (Eschenbach et al. 1991; Maliszewska-Kordybach et al. 2000; Klimkowicz-Pawlas and Maliszewska-Kordybach 2003). The inhibition of dehydrogenases and urease detected during this experiment might be the result of direct toxic effects of PAHs on the microorganisms, but might also be an indirect result of the deterioration of soil physical properties.

Among the invertebrates, earthworms are used for testing soil environmental pollution (ISO 15799: 2003). High sensitivity of earthworms to soil contaminants, including PAHs, is because of the fact that these organisms consume harmful compounds through the digestive tract, together with soil particles. Numerous studies have confirmed the negative effects of PAHs on the survivability and reproduction of earthworms (Brown et al. 2004; Eom et al. 2007). The present study confirmed the high sensitivity of earthworms to soil contamination with mineral oil containing PAHs. Both population density and biomass of these animals were strongly decreased when oil for lubrication of chainsaws was added to the soil, and the impact of the oil on biomass and quantity of earthworms was directly proportional to the amount of oil added.

A major problem when considering the adverse effects of exposure of soil organisms to PAHs is the duration of exposure to these substances and, in the case of model experiments, the period elapsed since PAHs were administered to the soil. Studies under laboratory conditions have shown that toxicity of various PAHs changes with time. The contributing factors for this phenomenon are the intensity of sorption of hydrocarbons by soil organic matter and regulation of their bioavailability, the mechanism of adaptation of soil organisms to the presence of PAHs in the soil (Lipińska et al. 2013), as well as the possibility of interactions between PAHs and other pollutants, especially heavy metals, whose intensity varies with time (Shen et al. 2006). In this experiment, the soil samples were of one soil type, with similar soil organic matter content. Soil samples were collected from areas directly adjacent to each other, so one can assume that the impact of this factor on modification of bioavailability of PAHs contained in oil would be limited. Concentrations of PAHs measured in the experiment did not exceed the permissible levels of these compounds in the soil as per the relevant soil quality standards (Trenck et al. 1994; Jensen and Folker-Hansen 1995). In this study, the cumulative content of the five PAHs measured (phenanthrene, pyrene, chrysene, benzo(a)pyrene, and dibenz[a,h]anthracene) did not exceed 0.3 mg kg^{-1} in the soil sample where the highest amount of mineral oil was administered. However, the adverse effect of oil contamination on enzymatic activity and earthworm biomass has been established. Soil enzyme activities have been considered useful parameters to provide a biological assessment of soil quality (Andreoni et al. 2004; Alrumman et al. 2015).

The activity of microorganisms and their enzymes is also determined by the properties of hydrocarbons, such as their degree of solubility in the soil solution and the amount of benzene rings in molecules of different hydrocarbons. Hydrocarbons with two, three, and four rings (e.g., naphthalene, phenanthrene, anthracene, and pyrene) are generally susceptible to microbial decomposition, and in the early stages of experiments might cause an increase in the activity of microorganisms and their enzymes (Smreczak and Maliszewska-Kordybach 2003; Zhon et al. 2010). Hydrocarbons with more than four benzene rings are strongly adsorbed and thus poorly bioavailable. In PAH degradation experiments in soil, a biphasic process can usually be observed; a rapid decline in contaminant content will occur in the first stage, while slow decomposition takes place in the second phase. The extended contact of PAHs with soil results in the formation of strong bonds with the soil components and penetration of these compounds into the soil micropores (Semple et al. 2003; Smreczak et al. 2013). In this experiment, the enzymatic activity and population and biomass of earthworms were measured 30 days after oil had been administered to the soil. Based on the results of model tests (Maliszewska-Kordybach et al. 2000), it was assumed that after this length of time the symptoms of the negative impacts of PAHs on soil biological parameters had stabilized.

5 CONCLUSIONS

Mechanization of forestry work leads to the emission of pollutants into the natural environment, including PAHs. In this study, the highest PAH concentrations were observed for chrysene and pyrene. Oil contamination modified the physical properties of forest soils. Oil at 100 g/m^2 caused a decrease of 4 % in air-filled porosity, and the highest amount of oil (200 g/m^2) resulted in a decrease in air-filled porosity of nearly 10 % compared with unpolluted soil. The addition of oil to soil resulted in the increased water resistance of different size aggregates; this influence was stronger the higher the amount of oil introduced into the soil. Enzyme activity was negatively correlated with PAH content. The lowest amount of oil (50 g/m^2) resulted in a slight decrease in urease activity, while activity was reduced by 40 and 50 % at 100 and 200 g/m^2, respectively, compared with the control. Earthworm population density reflected the state of oil-contaminated soils. The biomass of earthworms was directly proportional to the amount of oil applied.

ACKNOWLEDGMENTS

A subsidy for the maintenance; the scientific potential financed by the Polish MNiSW.

REFERENCES

1. Abosede, E. E. (2013). Effect of crude oil pollution on some soil physical properties. *Journal of Agriculture and Veterinary Science, 6*(3), 14–17.
2. Aislabie, J. M., Balks, M. R., Foght, J. M., & Waterhouse, E. J. (2004). Hydrocarbon spills on antarctic soils: effects and management. *Environmental Science & Technology, 38*(5), 1265–1274.
3. Alef, K., & Nannipieri, P. (1995). Enzyme activities. In K. Alef & P. Nannipieri (Eds.), *Methods in applied soil microbiology and biochemistry*. London: Academic Press.
4. Alrumman, S. A., Standing, D. B., & Paton, G. I. (2015). Effect of hydrocarbon contamination on soil microbial community and enzyme activity. *Journal of King Saud University—Science, 27*, 31–41.
5. Andreoni, V., Cavalca, L., Rao, M. A., Nocerino, G., Barnasconi, S., Dell'Amico, E., Colombo, M., & Gianfreda, L. (2004). Bacterial communities and enzyme activities of PAHs polluted soils. *Chemosphere, 57*, 401–412.
6. Bååth, E. (1989). Effects of heavy metals in soil on microbial processes and populations. *Water, Air, and Soil Polution, 47*, 335–379.
7. Baran, S., Bielińska, J. E., & Oleszczuk, P. (2004). Enzymatic activity in an airfield soil polluted with polycyclic aromatic hydrocarbons. *Geoderma, 118*, 221–232.
8. Brown, P. J., Long, S. M., Spurgeon, D. J., Svendsen, C., & Hankard, P. K. (2004). Toxicological and biochemical responses of the earthworm *Lumbricus rubellus* to pyrene, a non-carcinogenic polycyclic aromatic hydrocarbon. *Chemosphere, 57*, 1675–1681.

9. Clapperton M. J., Baker G. H., Fox C. A. 2007. Earthworms. In M. R. Carter, & E. G. Gregorich (Eds.), *Soil Sampling and Methods of Analysis*, Canadian Society of Soil Science (pp. 427--444).

10. Contreras-Ramos, S. M., Alvarez-Bernal, D., & Dendooven, L. (2006). Eisenia fetida increased removal of polycyclic aromatic hydrocarbons from soil. *Environmental Pollution, 141*, 396–401.

11. Dendooven, L., Alvarez-Bernal, D., & Contreras-Ramos, S. M. (2011). Earthworms, a means to accelerate removal of hydrocarbons (PAHs) from soil? A mini-review. *Pedobiologia, 54*, 187–192.

12. Eom, I. C., Rast, G., Veber, A. M., & Vasseur, P. (2007). Ecotoxicity of a polycyclic aromatic hydrocarbon (PAH)-contaminated soil. *Ecotoxicology and Enviromantal Safety, 67*, 190–205.

13. Eschenbach, A., Gehlen, P., & Bierl, R. (1991). Untersuchung zum Einfluss von Fluoranthene und Benzo(a)pyren auf Bodenmikroorganismen und zum mikrobiellen Abbau dieser Substanzen. *Mittteillungen d Dt Bodenkundliehen Gesellschaft, 63*, 91–94.

14. Giefing, D. F. (1991). Biooleje przeznaczone do smarowania układu tnącego pilarek. *Sylwan, 135*(12), 23–27.

15. ISO 15799 (2003). Soil quality—guidance on the ecotoxicological characterization of soils and soil materials.

16. Jensen, J., Folker-Hansen, P. (1995). Soil quality criteria for selected organic compounds. Danish Environmental Protection Agency. Working Report No. 47.

17. Khamehchiyan, M., Charkhabi, A. H., & Tajik, M. (2007). Effects of crude oil contamination on geotechnical properties of clayey and sandy soils. *Engineering Geology, 89*(3), 220–229.

18. Klimkowicz-Pawlas, A., & Maliszewska-Kordybach, B. (2003). Effect of anthracene and pyrene on dehydrogenases activity in soils exposed and unexposed to PAHs. *Water, Air, and Soil Pollution, 145*, 169–186.

19. Kotroczó, Z., Veres, Z., Fekete, I., Krakomperger, Z. S., Tóth, J. A., Lajtha, K., & Tóthmérész, B. (2014). Soil enzyme activity in response to long-

term organic matter manipulation. *Soil Biology and Biochemistry, 70*, 237–243.

20. Kucharski, J., Jastrzębska, E., Wyszkowska, J., & Hłasko, A. (2000). Wpływ zanieczyszczenia gleby olejem napędowym i benzyną ołowiową na jej aktywność enzymatyczną. *Zesz Probl Post Nauk Roln, 472*, 457–464.

21. Kucharski, J., Jastrzębska, E., & Wyszkowska, J. (2004). Wpływ substancji ropopochodnych na przebieg procesów amonifikacji i nitryfikacji. *Acta Agraria et Silvestria, Series Agraria, 42*, 249–255.

22. Kuyukina, M. S., Ivshina, I. B., Makarov, S. O., Litvinenko, L. V., Cunningham, C. J., & Philp, J. C. (2005). Effect of biosurfactants on crude oil desorption and mobilization in a soil system. *Environment International, 31*(2), 155–161.

23. Labud, V., Garcia, C., & Hernandez, T. (2007). Effect of hydrocarbon pollution on the microbial properties of a sandy and a clay soil. *Chemosphere, 66*(10), 1863–1871.

24. Lipińska, A., Kucharski, J., & Wyszkowska, J. (2013). Urease activity in soil contaminated with polycyclic aromatic hydrocarbons. *Polish Journal of Environmental Studies, 22*(5), 1393–1400.

25. Lipińska, A., Kucharski, J., & Wyszkowska, J. (2014). Activity of arylsulphatase in soil contaminated with polycyclic aromatic hydrocarbons. *Water, Air, and Soil Polution, 225*, 2097.

26. Maliszewska-Kordybach, B., Smreczak, B., & Martyniuk, S. (2000). Wpływ wielopierścieniowych węglowodorów aromatycznych na mikrobiologiczne właściwości gleb o zróżnicowanej kwasowości i zawartości substancji organicznych. *Soil Science Annual, 60*(3–4), 5–18.

27. Mueller, K. E., & Shann, J. R. (2006). Polycyclic aromatic hydrocarbons in spiked soil: impacts of bioavailability, microbial activity and trees. *Chemosphere, 64*, 1006–14.

28. Nam, J. J., Song, B. H., Eom, K. C., Lee, S. H., & Smith, A. (2003). Distribution of polycyclic aromatic hydrocarbons in agricultural soils in South Korea. *Chemosphere, 50*, 1281–9.

29. Ostrowska, A., Gawliński, S., & Szczubiałka, Z. (1991). *Metody analizy i oceny właściwości gleb i roślin* (p. 334). Warszawa: Katalog Instytutu Ochrony Środowiska.

30. Quyum, A., Achari, G., & Goodman, R. (2002). Effect of wetting and drying and dilution on moisture migration through oil contaminated hydrophobic soils. *Science of the Total Environment, 296*(1), 77–87.

31. Rudko, T., & Rybczyński, R. (2010). Właściwości smarne olejów roślinnych i mineralnych stosowanych w układach tnących pilarek. *Acta Agrophysica, 15*(1), 145–154.

32. Sannino, F., & Gianfreda, L. (2001). Pesticide influence on soil enzymatic activities. *Chemosphere, 45*, 417–425.

33. Semple, K. T., Morris, A. W., & Paton, G. I. (2003). Bioavailability of hydrophobic organic contaminants in soils: fundamental concepts and techniques for analysis. *European Journal of Soil Science, 54*, 809–818.

34. Shen, G., Lu, Y., & Hong, J. (2006). Combined effect of heavy metals and polycyclic aromatic hydrocarbons on urease activity in soil. *Ecotoxicology and Environment Safety, 63*, 474–480.

35. Smreczak, B., & Maliszewska-Kordybach, B. (2003). Preliminary studies on the evaluation of potentially bioavailable fractions of polycyclic aromatic hydrocarbons (PAHs) in soils contaminated with these compounds. *Archives of Environmental Protection, 29*(4), 41.

36. Smreczak, B., Klimkowicz-Pawlas, A., & Maliszewska-Kordybach, B. (2013). Biodostępność trwałych zanieczyszczeń organicznych (TZO) w glebach. *Studia i Raporty IUNG-PIB, 35*(9), 137–152.

37. Sutton, N. B., Maphosa, F., Morillo, J. A., Abu Al-Soud, W., Langenhoff, A. A., Grotenhuis, T., Rijnaarts, H. H., & Smidt, H. (2013). Impact of long-term diesel contamination on soil microbial community structure. *Applied and Environmental Microbiology, 79*(2), 619–630.

38. Swartz, R. C., Schults, D. W., Ozretich, R. J., Lamberson, J. O., Cole, F. A., DeWitt, T. H., Redmond, M. S., & Ferraro, S. P. (1995). PAH: a model to predict the toxicity of polynuclear aromatic hydrocarbon mixtures in field-collected sediments. *Environmental Toxicology & Chemistry, 11*, 1977–1987.

39. Townsend, G. T., Prince, R. C., & Suflita, J. M. (2003). Anaerobic oxidation of crude oil hydrocarbons by the resident microorganisms of a contaminated anoxic aquifer. *Environmental Science & Technology, 37*(22), 5213–5218. Trenck, K. T., Ruf, J., & Flittner, M. (1994). Guide values for contaminated sites. *Environmental Science and Pollution Research, 1*, 253–261.

40. Veres, Z., Kotroczó, Z., Fekete, I., Tóth, J. A., Lajtha, K., Townsend, K., & Tóthmérész, B. (2015). Soil extracellular enzyme activities are sensitive indicators of detrital inputs and carbon availability. *Applied Soil Ecology, 92*, 18–23.

41. Zhon, X., Wu, W., Zhon, L., Liong, J., & Jiang, T. (2010). Interactive effect of dissolved organic matter and phenanthrene on soil enzymatic activities. *Journal of Environmental Sciences, 22*(4), 607–614.

Chapter 4

Soil Dehydrogenases as an Indicator of Contamination of the Environment with Petroleum Products

Grażyna Kaczyńska, Agata Borowik, Jadwiga Wyszkowska

Department of Microbiology, University of Warmia and Mazury in Olsztyn

ABSTRACT

The aim of the research was to compare the effects of various petroleum products, biodiesel, diesel oil, fuel oil and unleaded petrol on soil dehydrogenases, and to evaluate biostimulation with compost and urea in the restoration of homeostasis of the soil contaminated with these products. The obtained results allowed for defining the weight of dehydrogenases in monitoring of the environment subjected to pressure from petroleum hydrocarbons. The studies were carried out under laboratory conditions for 180 days, and loamy sand was the soil formation used in the experiment. The petroleum products were used in the following amounts: 0, 2, 4, 8 and 16 g kg^{-1} DM of soil. Indices of the influence of the petroleum product and the stimulating substance on the activity of dehydrogenases were calculated. It was proved that the petroleum products affect soil dehydrogenases in various ways. Biodiesel, diesel oil and fuel oil stimulate these enzymes, while petrol acts as an inhibitor. Among the substances tested regarding biostimulation of soils contaminated with petroleum products, compost is definitely more useful than urea, and therefore, the former should be used for the remediation of such soils. Stimulation of dehydrogenases by compost, both in contaminated and non-contaminated soils, proves that it may accelerate microbiological degradation of petroleum-derived contaminants.

KEYWORDS

Soil dehydrogenases, Soil biostimulation, Petroleum products, Remediation

1 INTRODUCTION

Preservation of the biological equilibrium in the soil depends on numerous factors, which may be divided into chemical, physical and biological factors. The last group of parameters is particularly susceptible to modifications caused by any disturbance occurring in the soil and water environment. Enzymatic activity of the soil and proliferation of soil microorganisms are the best indicators of the stability and fertility of soil ecosystems. This is because of the immediate response of the biochemical activity of the soil to any disturbance of the environment. The soil environment is a source of an immense pool of enzymes. It includes representatives of every enzyme class, i.e. oxidoreductases, hydrolases, isomerases, ligases, liases and transferases. They all perform key functions in the process of the conversion of organic substances and energy (Gu et al. 2009). Soil dehydrogenases (EC 1.1.1.) are the main representatives of the oxidoreductase class.

Among all enzyme groups of the soil environment, dehydrogenases belong to one of the most important ones, because they occur in all living microorganism cells (Moeskops et al. 2010). As they are closely connected with microbiological redox processes (Gu et al. 2009), dehydrogenases are often considered as an index of the general microbiological activity of the soil (Moeskops et al. 2010). The significance of dehydrogenases as a pollution indicator is additionally supported by their lack of ability to accumulate in the extracellular environment. The role of dehydrogenases consists in the biological oxidation of organic matter in the soil by hydrogen transfer from the organic substrate to inorganic acceptors (Zhang et al. 2010). It should be emphasised that in the environment, many specific types of dehydrogenases occur, differing in terms of, among others, the coenzyme type (nicotinamide adenine dinucleotide, nicotinamide adenine dinucleotide phosphate or flavin adenine dinucleotide). Thanks to the ability to carry out dehydrogenation or hydrogenation, dehydrogenases are involved in the enzymatic systems of all living microorganisms (Subhani et al. 2001). Therefore, soil dehydrogenases are reported as an indicator of microbiological redox systems. Brzezińska et al. (2001) ascertained that these enzymes may use not only oxygen molecules as electron acceptors but also other compounds which occur in cells of anaerobic microorganisms. Thus, the activity of dehydrogenases reflects the rate of transformations occurring in the soil.

Petroleum products cause great devastation in soil and water ecosystems (Yeung et al. 2011). This results from the fact that these substances are mixtures of organic compounds with a low bioavailability, often described as potentially carcinogenic and mutagenic (Janbandhu and Fulekar 2011; Souza et al. 2014). Yeung et al. (2011) report that 1 dm^3 of oil pollutes 1 million dm^3 of water because of its ability to form a thin film with a surface area of 1000 m^2. This film prevents gas exchange and blocks the access of sunlight. Thus, it is not surprising that petroleum hydrocarbons belong to compounds considered highly toxic for soil ecosystems and human health (Albert and Tanee 2011).

In respect of chemical character, hydrocarbons may be included among bioresistant organic compounds (Bio-ROCs), which are able to completely or partially inhibit the metabolism of aerobic microorganisms. This is a consequence of the poor solubility of these substances in water, limited adsorbability and complex molecular structure (Semrany et al. 2012). As a consequence of their specific chemical features, such compounds are reduced by few enzymes (Lloret et al. 2010). As a result of the high potential of petroleum hydrocarbons to accumulate in the soil environment, together with their resistance to biodegradation, these compounds cause substantial changes in biological parameters, such as enzymatic activity and abundance of microbes (Pérez-Leblic et al. 2010, Zhan et al. 2010). One technique attempting to accelerate the natural degradation of petroleum products by the introduction of nutrients is called biostimulation (Chemlal et al.2012).

The aim of such actions consists in the stimulation of native microorganisms, which become resistant to the already occurring disturbances. Currently, it is a fast-growing research theme. This dependence is probably affected not only by the type of the soil (Aspray et al. 2008) or the kind of contaminant (Dilly et al. 2011) but also by the enzymatic system of the microbes. Lamy et al. (2013) suggest that the biostimulation is favourable only when, for the mineralisation of the compound used for it, the cell may utilise the enzymatic complexes of metabolic pathways used for degradation of the contaminating compounds.

The aim of the research was to compare the effects of various petroleum products, biodiesel, diesel oil, fuel oil and unleaded petrol on soil dehydrogenases, and to evaluate biostimulation with compost and urea in the restoration of the homeostasis of soil contaminated with these products. The obtained results allowed for defining the weight of dehydrogenases in the monitoring of environments subjected to pressure from petroleum hydrocarbons.

2 MATERIAL AND METHODS

2.1 Soil

In the research, soil from the Tomaszkowo Didactic and Experimental Station (NE part of Poland, 53.7161° N, 20.4167° E) was used. The soil material was collected from the topsoil of typical brown earths (Eutric Cambisol). According to the graining classification of the US Department of Agriculture, it was soil with a granulometric composition of loamy sand, with characteristics shown in Table 1.

Table 1. Physicochemical properties of the soil

Properties	Unit	Value
The percentage of the fraction (*d*)	$2.00 \geq d > 0.05$ mm	79.43
	$0.05 \geq d > 0.002$ mm	18.82
	$d \leq 0.002$ mm	1.75
pH_{KCl}		4.69
HAC	mM(+)kg^{-1} DM of soil	24.10
EBC		32.70
CEC		56.80
BS	%	57.57
C_{org}	g kg^{-1} DM of soil	11.60
N_c		1.39
K_a	mg kg^{-1} DM of soil	129.48
P_a		98.56
Mg_a		45.00

HAC hydrolytic acidity, *EBC* sum of exchangeable cations, *CEC* cation exchange capacity, *BS* base saturation, *org* organic, *c* total, *a* available

2.2 Petroleum Products

In the experiment, the following petroleum products were used: biodiesel, diesel oil, fuel oil and unleaded petrol. Their basic properties are shown in Table 2.

2.3 SUBSTANCES USED FOR BIOSTIMULATION

Compost and urea were used for the biostimulation of the soil contaminated with petroleum products. The compost was manufactured by a company NOLET (Poland) as a result of the aerobic composting of conifer sawdust and turkey litter. The organic carbon content in the compost amounted 402.0 g kg^{-1}, to total nitrogen 23.1 g kg^{-1}, available phosphorus 4.7 g kg^{-1}, available potassium 10.2 g kg^{-1} and available magnesium 2.2 g kg^{-1} and pH_{KCl} 7.08. The exchangeable cation contents were as follows: sodium 0.3 g kg^{-1}, potassium 0.8 g kg^{-1}, magnesium 0.5 g kg^{-1} and calcium 2.7 g kg^{-1}. Urea was manufactured by EUROCHEM BDG (Poland). Its properties are as follows: molar mass 60.06 g mol^{-1}, density 1.32 g $cm^{-3}$and solubility in water 480 g dm^{-3}.

2.4 Experimental Procedure

The experiments were carried out in triplicate under laboratory conditions, with the following variable factors: (a) type of petroleum product. biodiesel, diesel oil, fuel oil and unleaded petrol; (b) degree of contamination of the soil with the petroleum products, 0, 2, 4, 8 and 16 g kg^{-1} DM of soil; (c) type of the biostimulating substance, compost and urea, and (d) incubation time of the soil, 15, 30, 60, 90 and 180 days. One hundred grams of air-dried soil (loamy sand) screened through a sieve with a mesh size of 2 mm was weighed out to each of the beakers with a volume of 150 cm^3.

In order to neutralise acidification of the soil, it was limed using $CaCO_3$ before the experiment, in the amount of 0.18 g kg^{-1} DM of soil. In order to carry out the biostimulation of natural soil microbiota, compost in the amount of 10 g kg^{-1} DM of soil and urea in the amount of 250 mg N kg^{-1} DM of soil were added to the respective soil samples. Soil thoroughly mixed with the petroleum products and, in appropriate objects, with the compost or urea was brought to 50 % of maximum water capacity, protected and put in an incubator with a constant temperature 25 °C and no access to light. During the whole experiment cycle (180 days), the humidity level of the samples was checked once per week, and the water losses were supplemented.

Table 2. Properties of petroleum products

Parameter	Unit	Biodiesel (Bd) Value	Diesel (D)	Fuel oil (Fo)	Unleaded petrol (P)
Number of carbon atoms	—	C_6–C_{12}	C_9–C_{25}	C_9–C_{20}	C_5–C_{12}
Density	$g\ cm^{-3}$	0.88–0.89	0.82–0.85	0.80–0.91	0.72–0.78
Calorific value	$MJ\ kg^{-1}$	37–39	42–45	42–45	42–44
Cetane number	—	51	55	55	b.d.
PAH content	%	dl	11	dl	20
Content of solid impurities	$mg\ kg^{-1}$	24	24	24	dl
Sulphur content	$mg\ kg^{-1}$	10	10	2–3	10
Ignition point	°C	170	56	56–62	−51
Boiling point	°C	>350	175–180	180–360	30–210
Self-ignition point	°C	260	260	270	350
Acute toxicity by inhalation (DNEL)	$mg\ m^{-3}$ 15 min^{-1}	dl	4300	4300	2800

dl data lack

2.5 Determination of the Activity of Soil Dehydrogenases

After 15, 30, 60, 90 and 180 days, the experiments were liquidated and the activity of soil dehydrogenases was determined according to the procedure described in Öhlinger (1996). The determinations were carried out in triplicate. From each soil sample, 6 g of the soil was weighed out 0.06 g $CaCO_3$, and 1 cm^3 3 % aqueous solution of 1,3,5-phenyl-tetrazolium chloride and 2.5 cm^3 demineralised water were added to it successively. The samples were incubated for 24 h in 37 °C. After the incubation, 25 cm^3 methyl alcohol was added to every sample. Then, the content of each sample was thoroughly mixed and transferred quantitatively onto a filtering set consisting of a 50-cm^3 volumetric flask and a funnel with a quantitative hard filter with a diameter of 9 cm. This operation was repeated twice. After filtering, the flasks were made up to the mark with methyl alcohol, their contents were thoroughly mixed and absorbances were measured on a spectrophotometer Aquarius CE7500 Cecil Instruments, at a wavelength of $\lambda = 485$ nm. The results were presented as micromoles of triphenylformazan per kilogram DM per hour.

2.6 Determination of the Effect Indices of a Given Petroleum Product and Stimulating Substance

Based on the activity of soil dehydrogenases, indices of the petroleum product effect and stimulating substance effect were calculated using the following formulas:

$$\mathrm{IF}_{pp} = \frac{A_{pp}}{A_0}$$

Index of the petroleum product effect

A $_{pp}$ Activity of dehydrogenases in the soil contaminated with the petroleum product

A $_0$ Activity of dehydrogenases in the non-contaminated soil

$$\mathrm{IF}_b = \frac{A_b}{A}$$

Index of the stimulating substance effect

A $_b$ Activity of dehydrogenases in the soil subjected to biostimulation

A Activity of dehydrogenases in the soil not subjected to biostimulation

If IF = 1, there is no influence of the tested factor on dehydrogenases; IF < 1, there is inhibition of the dehydrogenases activity by the tested factor and IF > 1, there is stimulation of the dehydrogenases activity by the tested factor.

2.7 Statistical Analysis

The results of the studies were developed statistically using a statistical software package STATISTICA v. 12.0 (Statsoft, Inc., Statistica 2014). Based on the analysis of the effect measure η^2 by variance analysis—ANOVA—the percentage shares of all variable factors affecting the activity of dehydrogenases were defined. Homogeneous groups were calculated using Tukey's test with $P = 0.01$. The diversified influence of the petroleum products on the activity of dehydrogenases was illustrated using the index of the petroleum product effect and that of biostimulation with compost and urea—using the index of the stimulating substance effect.

3 RESULTS

3.1 Activity of Dehydrogenases

The activity of dehydrogenases in the soil non-contaminated with petroleum products was in the range of 3.967 μM TFF to 16.414 μM TFF kg^{-1} DM h^{-1}, depending on the incubation time (Table 3). In the soil fertilised with compost, it amounted from 9.845 μM TFF to 12.887 μM TFF, while in the soil fertilised with urea—from 0.676 μM TFF to 5.815 μM TFF. On average, irrespective of the test day, the activity in the soil fertilised with compost was 16 % higher, and in the case of urea—72 % lower than that of the non-fertilised soil.

Biodiesel (Bd) stimulated the activity of soil dehydrogenases on every test day (Table 4). Over 90 days, the activity decreased, together with an increase in the degree of contamination with Bd. On the 180th day, a decrease in the activity occurred in the soil subjected to the effect of 2 g Bd kg^{-1} DM, while doses of 4–16 g kg^{-1} increased the activity by from 0.660 to 0.799 times. The response of dehydrogenases to contamination with diesel oil (D) was also significant (Table 4). On almost all test days, the activity of these enzymes increased, together with an increase in the contamination degree. Such a tendency lasted for 180 days. However, the stimulating action of the lowest dose (2 g kg^{-1} DM of soil) declined on the 90th and 180th days.

Table 3. Activities of dehydrogenases in soil not contaminated with petroleum products (µmol TFF kg^{-1} DM of soil)

Addition	Incubation time, days				
	15	30	60	90	180
0	8.317b	11.720a	16.414a	3.967b	7.651b
Compost	12.887a	10.973b	9.845b	10.575a	11.474a
Urea	5.815c	0.938c	0.676c	2.408c	3.602c

The same letter means a homogeneous group in the columns

Fuel oil (Fo), similarly to biodiesel and diesel oil, stimulated dehydrogenases (Table 4). Its effect decreased more evidently the longer the experiment lasted. On the 180th day, the stimulating effect was exerted only by the dose of 16 g kg^{-1} DM of the soil. Petrol (P) affected dehydrogenases adversely. On almost all test days, the values of the index of the influence of this contaminant on dehydrogenases remained below 1 (Table 4). Values above 1 were observed only on the 60th day for doses of 4–16 g P kg^{-1} DM of the soil.

Summarising the evaluation of the influence of the tested substances on soil dehydrogenases, one may unequivocally ascertain that biodiesel, diesel oil and fuel oil stimulate these enzymes and petrol inhibits them (Fig. 1). Mean indices of the effect of the first three products were approximate and contained in the range from 1.639 (Fo) to 1.675 (Bd), while in case of petrol, the value of this index was significantly lower than 1 (0.717), proving its inhibitive influence on dehydrogenases (Fig. 2.)

Table 4. Index of the influence of petroleum products (IF_{pp}) on the activity of soil dehydrogenases

Dose Pp	Incubation time, days				
g kg^{-1} DM of soil	15	30	60	90	180
Biodiesel (Bd)					
2	3.337a	2.008a	1.416a	2.208a	0.886d
4	3.378a	1.982a	1.397a	1.836b	1.660b
8	1.997b	1.531b	1.290b	1.368c	1.763a
16	0.910c	0.933c	0.715c	1.089d	1.799a
Diesel (D)					
2	1.857c	1.533c	1.328b	0.828d	0.966c
4	2.278b	1.559bc	1.313b	1.510c	1.323b
8	2.587a	1.603b	1.484a	1.786b	1.324b
16	2.628a	1.668a	1.298b	2.028a	2.366a
Fuel oil (Fo)					
2	1.806d	1.333c	1.208c	0.742d	1.138c
4	2.221c	1.535b	1.248bc	1.224c	1.218b
8	2.461b	1.572b	1.287b	2.097b	1.459a
16	2.899a	1.819a	1.609a	2.417a	1.494a
Unleaded petrol (P)					
2	0.855a	0.915a	0.997d	0.426a	0.836a
4	0.640b	0.774b	1.102c	0.386b	0.622b
8	0.582c	0.698c	1.286b	0.336c	0.534c
16	0.588c	0.622c	1.374a	0.259d	0.500c

The same letter means a homogeneous group in the columns for a petroleum product (Pp)

3.2 Biostimulation of the Soil

The biostimulation of the soil with compost affected dehydrogenases favourably (Table 5). In the soil non-contaminated with petroleum products, the compost increased the activity by 45 % on average. The mean index of the effect amounted to 1.45. In the contaminated soils, mean indices calculated from all doses and dates amounted to in the objects from Bd 2.508, D 1.946, Fo 1.796 and P 2.724 (Fig. 3).

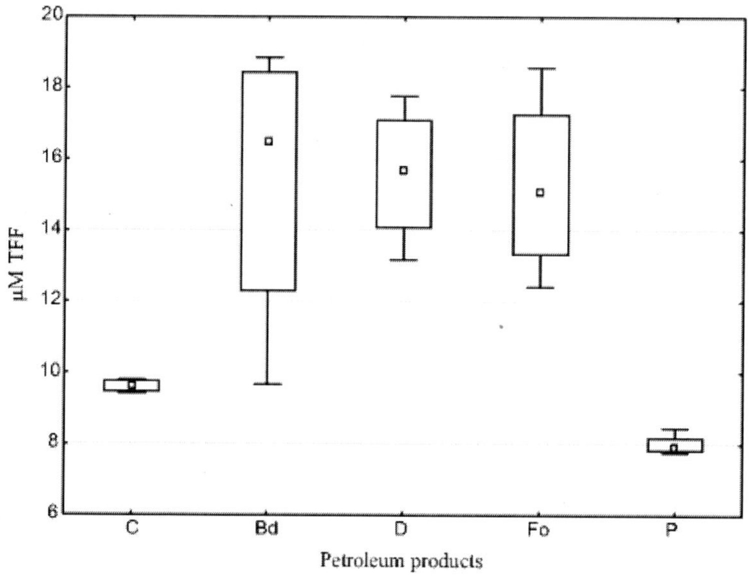

Figure. 1. Activity of dehydrogenases in soil contaminated with petroleum products (μM TFF kg^{-1} DM of soil). *C*control, *Bd* biodiesel, *D* diesel oil, *Fo* fuel oil, *P* unleaded petrol

The biostimulation of the soil with urea (Table 6) did not prove as effective as in the case of compost. In the non-contaminated soil, indices lower than 1 on all test days prove the negative influence of this fertiliser on dehydrogenases. In the contaminated soils, the effect of urea depended on the type of petroleum product. In the soil with biodiesel, the mean index of the urea effect amounted to 1.681, diesel oil 0.520, fuel oil 0.469 and unleaded petrol 1.165 (Fig. 3).

The data presented in Table 7 indicate unequivocally that the activity of dehydrogenases varied over time. Also, it depended on the biostimulation applied. The effects of biodiesel, diesel oil and fuel oil were mostly determined by the degree of soil contamination. The share of this factor in the formation of dehydrogenase activity ranged from 21.30 % (Fo) to 26.90 % (D). On the other hand, the dose of petrol did not affect dehydrogenases significantly, while compared to other variables, its effect was limited to 1.08 %. The biostimulation of the soil with compost and urea affected dehydrogenases significantly. The share of this factor in the formation of the activity ranged from 17.17 % (D) to 30.69 % (P) depending on the contamination type. Also, the incubation time of the soil affected these enzymes significantly. This independent variable determined the activity of dehydrogenases in the range from 14.39 % (D) to 23.31 % (P).

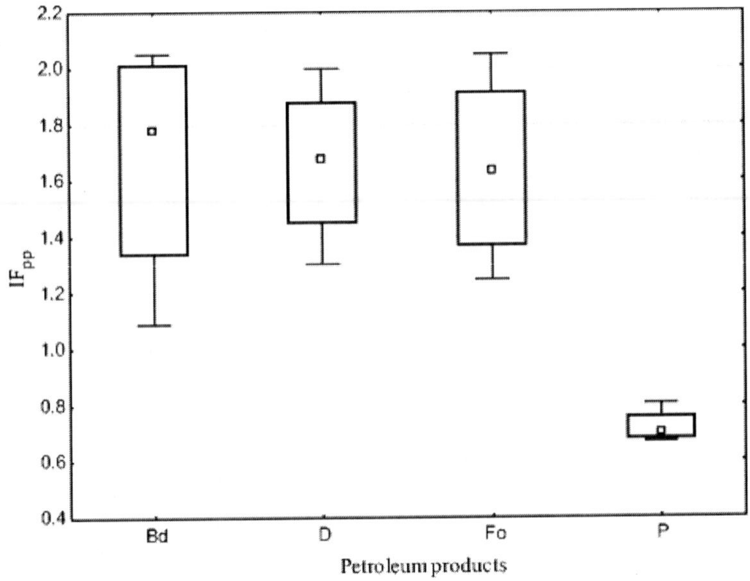

Figure. 2. Mean index of the influence of petroleum products (IF_{pp}) on the activity of soil dehydrogenases. *Bd*biodiesel, *D* diesel, *Fo* fuel oil, *P* unleaded petrol

4 DISCUSSION

4.1 Activity of Dehydrogenases

The activity of dehydrogenases is one of the most frequently used biological parameters for the evaluation of soil quality. Being intracellular enzymes, dehydrogenases are often mentioned as a good parameter for the estimation of contaminants (Galiulin et al. 2012; Lipińska et al. 2014). They perform as indicators of contamination of the environment with lipophilic substances because their presence reduces the redox potential as a result of the modification of oxygenation conditions, leading to a change in the terminal acceptor from oxygen molecule to Fe^{3+} cations. These are additional factors shaping the diversity of soil microorganisms. According to Galiulin et al. (2012), the degradation rate of hydrocarbons found in oil products increases with an increase in the activity of dehydrogenases which are directly involved in this process. Comparing the effect of diesel oil and biodiesel, the authors of Lapinskiené et al. (2006) demonstrated a statistically significant dependence between the increasing diesel oil dose and the reduction of dehydrogenase activity. They strove to prove unequivocally the toxicity of the tested petroleum products. In our own studies, the influence of petroleum products on the activity of dehydrogenases depended strictly on their type. The activity of the analysed oxidoreductases was positively correlated with the degree of contamination of the soil with biodiesel, diesel oil and fuel oil and negatively with unleaded petrol. The stimulating influence of diesel oil and fuel oil on the activity of dehydrogenases is indicated by the studies of Kucharski and Jastrzębska (2006) and Wyszkowska et al. (2002). The high activity of these enzymes may result from an intensified growth of microorganisms and their increased activity, because diesel oil may be a good nutrient for some microorganisms (Kucharski and Jastrzębska 2005; Wyszkowska and Kucharski2005). It is noteworthy that a mixture of PAH petroleum products affects the enzymes still less adversely than PAH penetrating the environment in their pure state, e.g. naphthalene, phenanthrene, anthracene and pyrene (Lipińska et al. 2014). The negative influence of polycyclic aromatic hydrocarbons on the activity of dehydrogenases was reported in Gianfreda et al. (2005). They proved this on the basis of studies on samples of soil exposed to contamination by crude oil for 50 years. This report emphasises the particular susceptibility of dehydrogenases to the presence of petroleum products in the soil. Our studies lasted for only 180 days, but even during this period, significant changes in the influence of the individual petroleum products on dehydrogenase activity could be observed. As the deposition of petroleum products was one time at the start of the experiment, the tendency of a reducing pressure of all petroleum products on dehydrogenases could be observed. This could result from a partial biodegradation of hydrocarbons by soil microbes (Xia et al. 2015). The possibility to stimulate indigenous microorganisms by some petroleum-derived contaminations is indicated by Wu et al. (2014).

4.2 Biostimulation of the Soil

Supplementation of the soil with organic matter, as a repository of an available pool of organic compounds, is key during the adaptation of the ecosystem to unfavourable environmental conditions. Stimulation of the biochemical activity by the application of various organic compounds is discussed in numerous scientific papers (Tejada et al. 2008; Wyszkowska and Wyszkowski 2010; Sayara et al. 2010; Galiulin et al. 2014). Their authors try to use various substances as stimulants for the indigenous microorganisms. The proposed solutions include the application of compost (Mehta et al. 2014; Wyszkowska and Wyszkowski 2006; Wyszkowski and Ziółkowska 2013), urea (Komilis et al. 2009; Wyszkowska et al. 2006), finely ground straw, sawdust (Wyszkowska et al. 2002), biological preparations (Galiulin and Galiulina 2015) and whey (Jonsson and Östberg 2011). In a study by Galiulin and Galiulina (2015), biocompost and biopreparations enabled almost complete degradation of hydrocarbons from crude oil and gas condensate. The effectiveness of remediation was evaluated based on, among others, dehydrogenase activity, and the research findings have been applied in practice (Patent of the Russian Federation, RU2387996 (C1)).

Table 5. Index of the effect of biostimulation with compost (IF_b) on the activity of soil dehydrogenases

Dose Pp	Incubation time, days				
g kg^{-1} DM of soil	15	30	60	90	180
Biodiesel (Bd)					
0	1.549c	0.936d	0.600e	2.666cd	1.500d
2	1.124d	0.870d	0.779d	1.909d	3.279b
4	1.308d	1.339c	1.065c	2.857c	1.954c
8	2.610b	2.043b	1.479b	4.762b	1.916c
16	6.026a	3.545a	3.177a	6.058a	2.058a
Diesel (D)					
0	1.549c	0.936c	0.600c	2.666d	1.500c
2	1.395d	0.788d	0.390d	3.081c	1.329d
4	1.301d	1.015c	0.460d	1.925e	1.011e
8	2.041b	2.093b	0.808b	3.837b	2.235a
16	2.945a	3.229a	1.928a	5.035a	2.067b
Fuel oil (Fo)					

0	1.549c	0.936c	0.600c	2.666d	1.500a
2	1.530c	0.946c	0.346d	3.312e	1.140b
4	1.313d	1.014c	0.639c	2.795e	1.043c
8	2.385b	2.081b	0.932b	3.458b	0.861d
16	3.159a	3.161a	1.188a	3.824a	0.787e
Unleaded petrol (P)					
0	1.549d	0.936d	0.600d	2.666e	1.500d
2	2.031c	1.535a	0.639d	3.620d	1.615c
4	2.623a	1.528a	0.687c	4.208c	1.890b
8	2.193b	1.468b	0.862a	5.269b	2.048a
16	2.129bc	1.257c	0.759b	16.058a	2.055a

The same letter means a homogeneous group in the columns for a petroleum product (Pp)

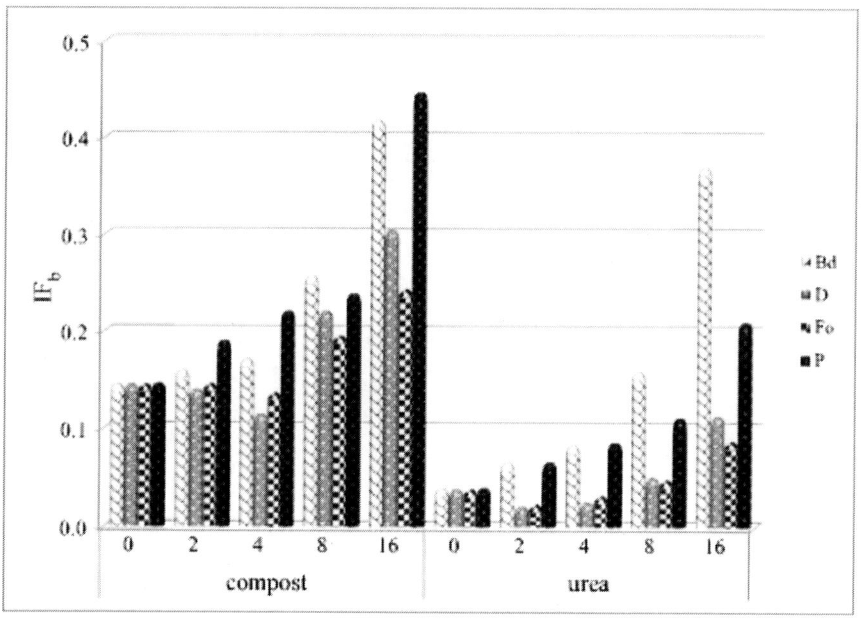

Figure. 3. Index of the influence of biostimulation (IF_b) with compost and urea on the activity of soil dehydrogenases 0, 2, 4, 8 and 16—dose of petroleum products (g kg^{-1} DM of soil). *Bd* biodiesel, *D* diesel, *Fo* fuel oil, *P*unleaded petrol

Table 6. Index of the effect of biostimulation with urea (IF_b) on the activity of soil dehydrogenases

Dose Pp	Incubation time, days				
g kg^{-1} DM of soil	15	30	60	90	180
Biodiesel (Bd)					
0	0.699e	0.080e	0.041c	0.607d	0.471c
2	0.917d	0.434d	0.012c	0.389e	1.492a
4	1.182c	0.803c	0.222b	1.089c	0.799b
8	2.678b	1.745b	0.279b	2.403b	0.770b
16	5.919a	3.838a	1.945a	5.228a	1.484a
Diesel (D)					
0	0.699cd	0.080b	0.041a	0.607b	0.471b
2	0.576d	0.029c	0.016c	0.125d	0.281d
4	0.743c	0.032c	0.018c	0.083d	0.373c
8	1.558b	0.066bc	0.019c	0.243c	0.621a
16	2.986a	0.617a	0.028b	1.602a	0.388c
Fuel oil (Fo)					
0	0.699d	0.080b	0.041a	0.607a	0.471c
2	0.680d	0.035c	0.015c	0.176c	0.214d
4	0.934c	0.039c	0.018bc	0.126cd	0.444c
8	1.469b	0.055bc	0.022b	0.075d	0.761b
16	2.566a	0.120a	0.019bc	0.288b	1.334a
Unleaded petrol (P)					
0	0.699e	0.080c	0.041b	0.607e	0.471b
2	0.925d	0.818b	0.030b	1.264c	0.182c
4	1.247c	0.896ab	0.054b	1.543bc	0.479b
8	1.397b	0.877ab	0.325a	1.961b	0.917a
16	1.522a	0.912a	0.399a	6.443a	1.102a

The same letter means a homogeneous group in the columns for a petroleum product (Pp)

Table 7. Participation of variable factors in the formation of dehydrogenases activity in percentage

Independent variable	Biodiesel (Bd)	Diesel (D)	Fuel oil (Fo)	Unleaded petrol (P)
Biostimulation (a)	17.17	20.55	18.95	30.69
Dose (b): Bd, D, Fo, P	25.30	26.90	21.30	1.08
Time (c)	20.69	14.39	18.13	23.31
Biostimulation·dose (ab)	16.00	11.57	8.30	2.56
Biostimulation·time (ac)	8.00	9.53	11.87	24.92
Dose·time (bc)	4.88	6.89	10.21	11.53
Biostimulation·dose·time (abc)	5.70	6.41	8.11	3.39
Error	0.03	0.03	1.85	1.54

The index representing the effect of stimulating substances (IF_b) on the activity of soil enzymes in soil contaminated with petroleum products was used in the study. The indices of biostimulation define the stability of the soil system and its production capacity. They also serve as indicators of emerging dangers in the soil. The discussed index clearly illustrates the tendencies of changes in the influence of the analysed substances on soil organisms. Thanks to a simple mathematical construction, there is a possibility to analyse the effects of the applied biostimulators and petroleum products. In our research, it was proved that compost was useful in the biostimulation of soils contaminated with petroleum products. It increased the activity of dehydrogenases both in contaminated and non-contaminated soils. Therefore, one may conclude that compost can accelerate the microbiological degradation of petroleum-derived contaminants, because it constitutes a source of necessary cosubstrates, nutrients or microorganisms, including the following: mesophilic and thermophilic bacteria, actinobacteria and fungi (Mehta et al. 2014). Galiulin and Galiulina (2015), who investigated the effects of biopreparations in soil contaminated with oil at 50 g kg^{-1}, reported a significant increase in dehydrogenase activity. In soil supplemented with Bioros (0.5 g kg^{-1}), which contains two groups of microorganisms: *Rodococcus* sp. and *Candida* sp. at 10^{10} cell g^{-1}, dehydrogenase activity was 19-fold higher than in oil-contaminated soil that was not supplemented with the above biopreparation. In soil treated with the Piksa biopreparation (100 g kg^{-1}), composed of a turf-manure mixture and enriched with hydrocarbonoxidising microorganisms in the quantity of 10^6 cells g^{-1}, dehydrogenase activity increased 11-fold. The noted increase in dehydrogenase activity after the application of biopreparations is indicative of microbial degradation of hydrocarbons from crude oil. This is confirmed by the studies of Xia et al. (2015), indicating that bioaugmentation connected with biostimulation yields a better effect in the degradation of hydrocarbons of petroleum products than bioaugmentation alone. Moreover, thanks to its structure, compost may improve the physical properties of the soil, changing its pH, water capacity and structure (Semple et al.2001). The favourable influence of compost on dehydrogenases activity observed in our studies is also confirmed by the investigations of Sayara et al. (2010), where the biochemical activity of soils contaminated with PAH was analysed.

In our studies, urea was decidedly less useful than compost in the biostimulation of soils contaminated with petroleum products. The index of the effect of biostimulation with urea (IF_b) on the activity of soil dehydrogenases in the non-contaminated soil and in the soil contaminated with fuel oil and diesel oil assumed values lower than 1, while in soil contaminated with biodiesel and petrol—values higher than 1. These low values of the IF_b indices may prove the lack of dependence between the attempt to compensate the C:N ratio and acceleration of the process of petroleum hydrocarbon degradation in the studied soil. Such a state may be a result of a change in the air-water conditions of the soil environment.

The small effect of nitrogen supplementation of the soil contaminated with petroleum-derived hydrocarbons is also proved by the studies Bento et al. (2005), Aspray et al. (2008) and Komilis et al. (2009). According to Lamy et al. (2013), the supplementation is favourable only when the cell can use its enzymes of the metabolic pathways of petroleum hydrocarbon degradation for decomposition of the stimulating substance. Also, the form of the nitrogen applied is of great importance. Inorganic nitrogen is better. According to Komilis et al. (2009), the application of nitrogen in the form of NH_4Cl is more favourable than the use of urea.

5 CONCLUSIONS

The petroleum products affect the soil dehydrogenases in various ways. Biodiesel, diesel oil and fuel oil stimulate these enzymes, while petrol acts as an inhibitor. Among the substances tested regarding biostimulation of soils contaminated with petroleum products, compost is definitely more useful than urea, and therefore, the former should be used for the remediation of such soils. Stimulation of dehydrogenases by compost, both in contaminated and non-contaminated soils, proves that it can accelerate the microbiological degradation of petroleum-derived contaminants.

REFERENCES

1. Albert, E., & Tanee, F. B. G. (2011). A laboratory trial of bioaugmentation for removal of total petroleum hydrocarbon (TPH) in Niger Delta soil using *Oscillatoria bornettia. J Microbiol Biotechnol, 1,* 147–168.

2. Aspray, T., Gluszek, A., & Carvalho, D. (2008). Effect of nitrogen amendment on respiration and respiratory quotient (RQ) in three hydrocarbon contaminated soils of different type. *Chemosphere, 72,* 947–951.

3. Bento, F. M., Camargo, F. A. O., Okeke, B. C., & Frankenberger, W. T. (2005). Comparative bioremediation of soils contaminated with diesel oil by natural attenuation, biostimulation and bioaugmentation. *Bioresour Technol, 96*, 1049–1055.

4. Brzezińska, M., Stępniewski, W., Stępniewska, Z., & Przywara, G. (2001). Effect of oxygen deficiency on soil dehydrogenase activity in a pot experiment with Triticale cv. Jago vegetation. *Int Agrophys, 15,* 145–149.

5. Chemlal, R., Tassist, A., Drouiche, M., Lounici, H., Drouiche, N., & Mameri, N. (2012). Microbiological aspects study of bioremediation of diesel contaminated soils by biopile technique. *Int Biodeter Biodegr, 75,* 201–206.

6. Dilly, O., Nii-Annang, S., Franke, G., Fischer, T., Buegger, F., & Zyakun, A. (2011). Resilience of microbial respiration, respiratory quotient and stable isotope characteristics to soil hydrocarbon addition. *Soil Biol Biochem, 43*, 1808–1811.

7. Galiulin, R. V., & Galiulina, R. A. (2015). Remediation of Polar Ecosystems Polluted by Gas Condensate and Oil Hydrocarbons by Biological Preparations. *The Open Ecology Journal, 8*(Suppl 1-M5), 40–43.

8. Galiulin, R. V., Bashkin, V. N., & Galiulina, R. A. (2012). Degradation of petroleum hydrocarbons in soil under the action of peat compost. *Solid Fuel Chem, 46*(5), 328–329.

9. Galiulin, R. V., Galiulina, R. A., & Bashkin, V. N. (2014). Reclamation of soil polluted by gas condensate using peat compost. *Solid Fuel Chem, 48*(5), 320–322.

10. Gianfreda, L., Rao, M. A., Piotrowska, A., Palumbo, G., & Colombo, C. (2005). Soil enzyme activities as affected by anthropogenic alterations: intensive agricultural practices and organic pollution. *Sci Total Environ, 341*, 265–279.

11. Gu, Y., Wag, P., & Kong, C. (2009). Urease, invertase, dehydrogenase and polyphenoloxidase activities in paddy soils influenced by allelophatic rice variety. *Eur J Soil Biol, 45*, 436–441.

12. Janbandhu, A., & Fulekar, M. H. (2011). Biodegradation of phenanthrene using adapter microbial consortium isolated from petrochemical contaminated environment. *J Hazard Mater, 187*, 333–340.

13. Jonsson, A. P., & Östberg, T. L. (2011). The effects of carbon sources and micronutrients in whey and fermented whey on the kinetics of phenanthrene biodegradation in diesel contaminated soil. *J Hazard Mater, 192*, 1171–1177.

14. Komilis, D., Vrohidou, A. E., & Voudrias, E. (2009). Kinetics of aerobic bioremediation of a diesel-contaminated sandy soil: effect of nitrogen addition. *Water Air Soil Pollut, 208*, 193–208.

15. Kucharski, J., & Jastrzębska, E. (2005). Effects of heating oil on the count of microorganisms and physico-chemical properties of soil. *Pol J Environ Stud, 14*(2), 195–204.

16. Kucharski, J., & Jastrzębska, E. (2006). Effect of heating oil on the activity of soil enzymes and the yield of yellow lupine. *Plant Soil Environ, 52*(5), 220–226.

17. Lamy, E., Tran, T. C., Mottelet, S., Pauss, A., & Schoefs, O. (2013). Relationships of respiratory quotient to microbial biomass and hydrocarbon contaminant degradation during soil bioremediation. *Int Biodeter Biodegr, 83*, 85–91.

18. Lapinskienė, A., Martinkus, P., & Rėbždaitė, V. (2006). Eco-toxicological studies of diesel and biodiesel fuels in aerated soil. *Environ Pollut, 142*, 432–437.

19. Lipińska, A., Kucharski, J., & Wyszkowska, J. (2014). The effect of polycyclic aromatic hydrocarbons on the structure of organotrophic bacteria and dehydrogenase activity in soil. *Polycyclic Aromatic Compounds, 34*(1), 35–53.

20. Lloret, L., Eibes, G., Lu-Chau, T. M., Moreira, M. T., Feijoo, G., & Lema, J. M. (2010). Laccase catalyzed degradation of anti-inflammatories and estrogens. *Biochem Eng J, 51*, 124–131.

21. Mehta, C. M., Uma Palni, U., Franke-Whittle, I. H., & Sharma, A. K. (2014). Compost: its role, mechanism and impact on reducing soil-borne plant diseases. *Waste Manag, 34*, 607–622.

22. Moeskops, B., Buchan, D., Sleutel, S., Herawaty, L., Husen, E., Saraswati, R., Setyorini, D., & De Neve, S. (2010). Soil microbial communities and activities under intensive organic and conventional vegetable farming in West Java, Indonesia. *Appl Soil Ecol, 45*, 112–120.

23. Öhlinger, R. (1996). Dehydrogenase activity with the substrate TTC. In F. Schinner, R. Ohlinger, E. Kandler, & R. Margesin (Eds.), *Methods in soil biology* (pp. 241–243). Berlin: Springer Verlag.

24. Patent of the Russian Federation, RU2387996 (C1)—2010-04-27 method of monitoring cleaning of soil contaminated with hydrocarbons and neutralisation of hydrocarbon sludge through analysis of dehydrogenase activity.

25. Pérez-Leblic, M. I., Turmero, A., Hernandez, M., Hernandez, A. J., Pastor, J., Ball, A. S., Rodrigues, J., & Arias, M. E. (2010). Influence of xenobiotic contaminants on landfill soil microbial activity and diversity. *J Environ Manage, 30*, 1–6.

26. Sayara, T., Sarra, M., & Sanchez, A. (2010). Effects of compost stability and contaminant concentration on the bioremediation of PAHs-contaminated soil through composting. *J Hazard Mater, 179*, 999–1006.

27. Semple, K. T., Reid, B. J., & Fermor, T. R. (2001). Impact of composting strategies on the treatment of soils contaminated with organic pollutants. *Environ Pollut, 112*, 269–283.

28. Semrany, S., Faviera, L., Djelal, H., Tahac, S., & Amranea, A. (2012). Bioaugmentation: possible solution in the treatment of bio-refractory organic compounds (Bio-ROCs). *Biochem Eng J, 69*, 75–86.

29. Souza, E. C., Vessoni-Penna, T. C., & de Souza Oliveira, R. P. (2014). Biosurfactant-enhanced hydrocarbon bioremediation: an overview. *Int Biodeter Biodegr, 89*, 88–94.

30. Statsoft, Inc., Statistica. (2014). Data analysis software system, version 12.0.

31. Subhani, A., Changyong, H., Zhengmiao, Y., Min, L., & El-ghamry, A. (2001). Impact of soil environment and agronomic practices on microbial/dehydrogenase enzyme activity in soil. A review. *Pak J Biol Sci, 4*, 333–338

32. Tejada, M., Gonzalez, J. L., Hernandez, M. T., & Garcia, C. (2008). Application of different organic amendments in a gasoline contaminated soil: effect on soil microbial properties. *Bioresour Technol, 99*, 2872–2880.

33. Wu, B., Lan, T., Lu, D., & Liu, Z. (2014). Ecological and enzymatic responses to petroleum contamination.*Environ Sci Processes Impacts, 16*, 1501–1509.

34. Wyszkowska, J., & Kucharski, J. (2005). Correlation between the number of cultivable microorganisms and soil contamination with diesel oil. *Pol J Environ Stud, 14*, 347–356.

35. Wyszkowska, J., & Wyszkowski, M. (2006). Role of compost, bentonite and lime in recovering the biochemical equilibrium of diesel oil contaminated soil. *Plant Soil Environ, 52*, 505–514.

36. Wyszkowska, J., & Wyszkowski, M. (2010). Activity of soil dehydrogenases, urease and acid and alkaline phosphatases in soil polluted with petroleum. *J Toxicol Environ Health, 73*, 1202–1210.

37. Wyszkowska, J., Kucharski, J., & Wałdowska, E. (2002). The influence of diesel oil contamination on soil enzymes activity. *Rostlinna Vyroba, 48,* 58–62.

38. Wyszkowska, J., Kucharski, M., & Kucharski, J. (2006). Application of the activity of soil enzymes in the evaluation of soil contamination by diesel oil. *Pol J Environ Stud, 3,* 501–506.

39. Wyszkowski, M., & Ziółkowska, A. (2013). Content of polycyclic aromatic hydrocarbons in soils polluted with petrol and diesel oil after remediation with plants and various substances. *Plant Soil Environ, 59,* 287–294.

40. Xia, W., Xia, J., Lia, J., Zhanga, D., Zhoub, Q., & Wangb, X. (2015). Studies on crude oil removal from pebbles by the application of biodiesel. *Mar Pollut Bull, 91,* 288–294.

41. Yeung, C. W., Law, B. A., Milligan, T. G., Lee, K., Whyte, L. G., & Greer, C. W. (2011). Analysis of bacterial diversity and metals in produced water, seawater and sediments from an offshore oil and gas production platform. *Mar Pollut Bull, 62,* 2095–2105

42. Zhan, X., Wu, W., Zhou, L., Liang, J., & Jiang, T. (2010). Interactive effect of dissolved organic matter and phenanthrene on soil enzymatic activities. *J Environ Sci, 22,* 607–614.

43. Zhang, N., He, X., Gao, Y., Li, Y., Wang, H., Ma, D., Zhang, R., & Yang, S. (2010). Pedogenic carbonate and soil dehydrogenase activity in response to soil organic matter in *Artemisia ordosica* community. *Pedosphere, 20,* 229–235.

Chapter 5

Chronic Toxicity of Weathered Oil-Contaminated Soil

Dinora Vázquez-Luna[1]

[1] Universidad Veracruzana, México

1. INTRODUCTION

Currently, the activities derived from the oil industry, such as extraction, transportation and processing of oil, have affected natural resources [1-3]. For decades, tropical lands have been contaminated by chronic oil spills, causing significant changes in physical and chemical characteristics of the soil, affecting plant development and reducing the growth of microorganisms [4]. Moreover, oil weathering and adaptation of some plants may hide the toxicity of high molecular weight hydrocarbons to other organisms. For this reason, it is necessary to incorporate aspects of *chronic toxicity of weathered oil-contaminated soil* in the study of "*Soil Pollution*". This chapter aims to examine the chronic effects of old spills on soil, plants and beneficial microbes, in order to support the creation of new remediation technologies, focused on face the new challenges of soil contamination.

Oil pollution is a global problem of increasing importance [5], is estimated that every year numerous spills affecting natural resources of Southeast Mexico, in 2011 were contaminated more than 2,063 hectares as a result of 217 oil spills that affected soil and sea; of which 85 were caused by uncontrolled illegal connections, number which increased 204% from 2009-2011 [6]. With respect to the damages to the ground, in 2012 reported an increase of 87% of leaks and spills, its main causes were corrosion damage and failure of materials in pipelines, in that year 30.07 hectares were contaminated in the Southern Region, joined this, there is lag in the care of cases from previous years. This year, the oil industry closed its operations with a total of 163.63 hectares waiting to be remedied, of which, 39.5% are located in the southeast [7].

The environmental problems caused by oil spills, is not limited to visible pollution, because there are chronic effects that silently endanger ecosystems, biodiversity and environmental balance, due to bioaccumulation, leaching and extension of contaminants into groundwater with potential effects on all living organisms [8]. In Mexico, current environmental regulations is the NOM-138-SEMARNAT/SS-2003 which establishes the maximum permissible limits of hydrocarbons in soil and specifications for its characterization and remediation [9], although in this standard are omitted criteria for assessing chronic effects on soil microorganisms and plants as a result of weathered oil contamination. However, the contributions of several studies show that pollution and waste generation caused by oil activities have deteriorated soil quality [10,11]; ie, the social responsibility of the industry and the government should not only attend acute environmental emergencies [12].

Crude oil is a complex mixture of thousands of compounds that when released to the environment, is subjected to physical, chemical and biological processes called weatherability [13]. This process includes adsorption, volatilization, dissolution, biodegradation, photolysis, oxidation, and hydrolysis. The effects of weathering are difficult to predict because it depends on many biotic and abiotic factors [14]. Therefore, the mobility of hydrocarbons is also influenced by natural factors [15], which involve: a) chemical processes (hydrolysis, oxidation, reduction, photolysis), b) transport and physical processes (adsorption, advection, dispersion, diffusion, volatilization and dissolution), and c) biological processes (biodegradation, metabolism and toxicity) [16, 17].

When an oil spill occurs, it covers the soil surface, but initially at high viscosity prevents penetration towards the subsoil. The oil is retained in the topsoil, during this phase, the light fraction is photo-oxidized and volatilized through the soil pore space and transported to the atmosphere, in this process involves the first n-alkanes (methane, ethane, propane and butane), which are evaporated in less than 24 hours in tropical climates [18]. The hydrocarbons which are not evaporated are incorporated into the soil to form a waterproofing layer that prevents the normal flow of water. This layer or film affects the structure, porosity, absorption and water penetration into the soil [19].

Subsequently, the oil soluble fraction diffuses into the soil solution through infiltration. The behavior of this fraction in the soil depends on the type of texture. The presence of fine texture allows the volatilization of some compounds (C5-C7), but when the texture is coarse, it can leach out and transport themselves to the groundwater, affecting other organisms [20], including the human [21-23]. Otherwise, clay soils rich in organic matter immobilize some compounds, reducing their toxicity and decreasing its spread and leaching through the soil profile [24, 25].

The most stable fraction of crude oil is composed with more than 18 carbons (Polycyclic Aromatic Hydrocarbons [PAHs] and polar compounds), which are adhered on the soil matrix decreasing the solubility and volatility [26], and increasing the capacity of adsorption on the mineral and organic fractions, owing to the high content of active surface of the soil (clay 2:1) and to the high molecular weight hydrocarbons [27]. At this stage, the development of bacteria and fungi have influence on mineralization as part of natural attenuation process, but it is not sufficient to remove all hydrocarbons weathered [28].

Recent studies indicate that the major impediments to the biodegradation of hydrocarbons are the physical and chemical properties of the soil, the degree of contamination and the molecular weight of the compounds (C10-C40), but with Enhanced Natural Attenuation (ENA) may be observed a biodegradation of 26.4% [29] up to 60% with enriched amendments, although the n-alkanes are not removed completely [30]. Therefore, it is important to study the toxic potential, because there are reports indicating that can be bioaccumulated PAHs in vegetables such as *C. pepo sp.* [31].

2. MATERIALS AND METHODS

In southeastern Mexico, the land has been affected as a result of extraction; handling and transportation of oil, there are also zones of oil discharges, which are deposited in the open, without any environmental protection measure [32]. The bad condition of the pipelines and the dispersion through surface runoff of rainwater, resulting from weather conditions become more complicated to calculate the chronic effects of soil contamination by hydrocarbons [10]. Accordingly, this study was conducted in three stages, which were analyzed the chemical and physical properties of the soil, the toxic effects of weathered hydrocarbons on the growth and development of seedlings and after 150 and 240 days of exposure, and finally was evaluated the behavior of beneficial soil microorganisms in rhizospheric soil.

2.1. Effect On The Physical And Chemical Properties Of The Soil

The study was undertaken at the facilities the *Colegio de Postgraduados*, located in Tabasco, Mexico. The soil with weathered oil was collected within 2 km of the Petrochemical "La Venta" (18 ° 04 '54 "N and 94 ° 02' 31" W), Figure 1. The pollution-free soil according to NOM-138-SEMARNAT/SS-2003 was located in the community of Santa Teresa Arroyo Hondo. The objective was to identify the types of soils and their level of similarity.

For both soils were determined the content of organic matter (Walkley and Black), pH (potentiometry), P and K exchangeable (by extraction with 1N ammonium acetate pH 7, quantification by atomic absorption and emission respectively), CEC (extraction 1N ammonium acetate pH 7, quantification by distillation and titration) and texture (Bouyoucos). The analytical methods used were those indicated in NOM-021-2000-RECNAT [33]. Total Petroleum Hydrocarbons (TPH) were determined by the technique reported in the NOM-138-SEMARNAT/SS-2003 [9].

2.2. Effects On Plants

Bioassays to determine the phytotoxicity of weathered oil on two species of legumes (*Crotalariaincana* and *Leucaena leucocephala*) were carried out in the facilities of the Laboratory of Soil Microbiology, Colegio de Postgraduados, Tabasco, Mexico. Two states were studied phenological (plant and seedling), in two treatments: a) soil with 150 $mg.kg^{-1}$ TPH (control treatment) and b) soil with 79.457 $mg.kg^{-1}$ TPH weathered (this has been contaminated by over 25 years). Both soils were characterized as Gleysols, with the same pedogenetic origin, as described in the previous section.

Bioassays were established under a Completely Randomized Design (CRD) with three replications and two legumes, these were selected to be species that grow wild in oiled areas, but the former has tolerance, while the second shows sensitivity to high concentrations of crude oil [34]. In each bioassay was used 208 Protocol of the Organization for Economic Cooperation and Development (OECD) modified according to [35], which allows easily identify the symptoms of stress in the plant.

Seedling bioassays: 50 seeds were sown *C. incana* and 25 *L. leucocephala* by repetition, respectively. Glass containers were used (32 x 22 x 5.5 cm). The number of seeds sown per plant species was calculated according to the size of the seed [36], the viability of the species [37] and the area of the container. Seeds were previously scarified to remove impermeable integument, which constitute a barrier for germination [38]. Scarification consisted of immersing the seeds in sulfuric acid for 15 minutes and washed with tap water subsequent to remove all acid residues [39]. Germination tests were performed to determine the initial seed viability, finding viability standard values in both blocks [40]. The test lasted 30 days and the variables evaluated were mortality, height, root length and dry matter accumulation aerial and root.

Plant bioassays: Bioassays were established plants seedlings 30 days from uncontaminated soil. Subsequently, two of these were transplanted into containers, 15 days after a plant was removed from each container. Exposure of plants to pollutant lasted 150 days to *C. incana* and 240 days for *L. leucocephala*, because of its tolerance respective. The physiological variables were evaluated: height, root length, biomass (leaves and stems), root biomass and number of nodes, leaves and fruits. During both assays was provided with water to field capacity and were not supplied nutrients (N, P and K) to avoid interference on the growth of the specimens. The material was weighed on an analytical balance to obtain the values of DM. The multiple comparison of means was performed by Tukey test (a = 0.05). The numerical results were analyzed with SAS software version 9.1, using PROC GLM.

2.3. Effect On Soil Microorganisms

Quantification of microorganisms was determined by viable count method for serial dilutions [41] for*Rhizobium* extracts in nodules and Free Living Nitrogen Fixing Bacteria (FLNFB) in rhizospheric soil. Culture media were used combined carbon and yeast-mannitol agar [42].

3. RESULTS AND DISCUSSION

3.1. Effect On The Physical And Chemical Properties Of The Soil

The soil profile and laboratory results confirmed that both soils are Gleysols, with the same pedogenetic origin [43] (Table 1), but the contamination had abnormalities in their chemical characteristics (Table 2), the which have been reported in other studies undertaken in the South East (Table 3).

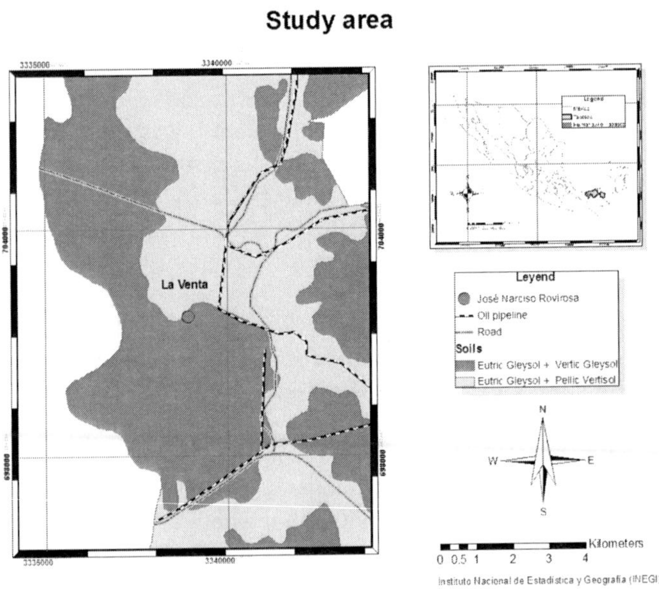

Figure 1. Study zone

Also, many physical properties of soil are altered, such as water retention capacity; this is because when an oil spill occurs, hydrocarbons bit compete with water filling the pores [49]. On the other hand, increasing the moisture content in the soil reduces the adsorption of liquid hydrocarbons in the organic matter and clay and the filling of the pores and capillaries (Figure 2) [50].

The oil can form macroaggregates and macropores that increase water flow (Figure 3), and these changes inhibit the water retention in the soil, which is moistened only after a long period of contact [51], so that the plants may suffer water stress and die.

3.2. Effects On Plants

The results indicated that the lower height and root length were found in soil with 79,457 $mg.kg^{-1}$ of weathered oil, this can be explained because of the limited development of these. Both legumes formed less biomass in soils with higher concentrations of oil, which was associated with lower production of biomass in leaves and stems, as a result of the presence of the contaminant in the soil (Table 4 and 5), which limits the entry of water to the plant. However, the hydrophobic effect also affects the early stages of germination as evidenced with a delay of up to five days in the emergency [52], even in high concentrations of TPH, germination may be inhibited completely [53].

Table 1. Chemical properties and concentration of TPH in the soils studied.

Sampling 0-30 cm	pH (H_2O)	OM (%)	P mg.kg^{-1}	K Cmol(+)kg^{-1}	CEC	Clay	Silt (%)	Sand	Textural classification	TPH (mg.kg^{-1})
Soil †	6.3	10.2	23.1	0.35	45.2	61	21	18	Clayey	150
Soil ††	4.2	25.8	3.58	0.4	43.5	48	33	19	Clayey	79,457

* pH 1:2 (potentiometry), organic matter (Walkley and Black), P and K (extraction with 1N ammonium acetate pH 7 and quantification by atomic absorption and emission), CEC (Cation Exchange Capacity) (extraction with 1N ammonium acetate and quantification by distillation and titration), texture (Bouyoucos).

† Control soil

†† Soil with weathered oil (OM without previous removal of TPH)

Table 2. Interpretation of the characteristics of the soils studied. The interpretation was based on the ranges indicated in NOM-021-2000-RECNAT.

Sampling 0-30cm	pH	OM (%)	P mg.kg^{-1}	K	CEC C mol(+) kg^{-1}
Control soil	Moderately acid	Medium	High	Medium	Very High
Polluted soil ††	Strongly acidic	Very high	Low	Medium	Very High

†† Soil with weathered oil (OM without previous removal of TPH)

Table 3. Alterations reported in soils contaminated with TPH, in southeastern Mexico.

Alterations	**Effects**
High ratios of C / N and C / P	Unfavorable microbial growth [44]
Retaining TPH fractions in organic matter	Alter the solubility of phosphorus [45]
Increasing Na	Limitations in the production of plants [11, 46]
Decreasing pH	Decrease microbionas populations [47]
The electrical conductivity can be increased up to 5.6 times	Salinity [48]
Interruption in the interaction between cations Ca and K	Reduction in capacity of soil to retain Ca and K [32]

Figure 2. Alteration in the retaining moisture of soil.

Figure 3. Formation of macroaggregates in soil with 12,155 mg.kg^{-1} TPH weathered.

Some authors found that exposure to concentrations of 2 791, 9 025 and 79 457 mg.kg^{-1} of petroleum hydrocarbons in soil inhibited the vegetative growth and reduced plant biomass in seedlings of*Echinochloa polystachya*, *Brachiaria mutica* and *Cyperus sp* [10]. Others establish increased toxicity on the dry weight of rice seedlings (*Oryza sativa*) after 25 days of exposure to 90,000 mg.kg^{-1}weathered oil [54]. Biomass reduction is possibly due to widespread damage, which begins in the root system, hindering vegetative growth and therefore the accumulation of plant biomass. However, the first studies mentions that at low concentrations, oil could stimulate vegetative growth [55].

Crotalaria incana seedlings did not form nodules in soil contaminated with 79,457 mg.kg^{-1} of weathered oil after 30 days of exposure to oil. These results were similar to those obtained according to [34], who found that nodulation was completely inhibited in *Crotalaria sp.* and *Mimosa pigra* by concentrations above 50,000 $mg.kg^{-1}$.

In both plant species, growth decreased with higher concentration of hydrocarbons in soil (Table 5 and6). This response may be related to decreased water absorption through the roots for the presence of the hydrophobic film formed by the oil added to the soil [56]. The lack of water absorbed decreases cell turgor, reduces or inhibits the processes of incorporation of nutrients and also affects vegetative growth [57, 58]. Water stress is related to the water potential gradient (which depends on the conditions present in the soil) and the membrane permeability to water (which depends on the species [59] therefore, *Leucaena leucocephala* shown to be more tolerant to water deficit).

Crotaria incana plants showed a greater effect on biomass in response with exposure to the weathered oil (Figure 4), which may be related to increased toxicity of recalcitrants compounds. There are also some soil properties which allow the adsorption of pollutants [60]. Clay soils with high organic content and low pH may favor the persistence of toxic substances in the soil for a long time after the oil spill occurred [61, 34], due to the adhesiveness of organic matter [62].

Furthermore, fine texture allows the oil form a coarse structure on the outside and around the conglomerate making it waterproof [63], this has effects on root development, growth, and as a result will cause decrease in the accumulation biomass [59]. On the other hand, soil contamination by hydrocarbons can also modify some characteristics such as texture, bulk density, ratio of the particle size of the soil, reducing aeration and affecting the productive development of plants [64, 65].

3.3. Effect On Soil Microorganisms

In plants *Leucaena leucocephala* high concentration of weathered oil did not affect populations of*Rhizobium* into nodules, as happened in the case of *C. incana*, in which both populations were significantly lower in a shorter time exposure (Figure 5). This is because the oil alters the physical and chemical characteristics of the soil, causing the blockage of gas exchange with the atmosphere and affecting microbial populations. Furthermore, the weathered oil is adsorbed in the ground, being less accessible and more difficult to degrade by microorganisms [66, 67]. This may bring a direct impact on rhizobia, because they are aerobic bacteria that remain in the soil as saprophytes, until they infect a radical hair.

Table 4. Response of *Crotalaria incana* seedlings to 30 days of weathered oil exposure.

Species	Concentration	Height	Root length	Mortality	Biomass				
					Leaves	Stems	Aerial	Root	Total
	$mg.kg^{-1}$	cm		(%)	gr				
Leucaena	150	7.5a	19.0a	0.0a	3.3a	1.3a	4.6a	1.1a	5.7a
	79,457	5.1b	5.1b	4.2a	1.5b	0.5b	2.0b	0.5b	2.5b

Values with different letter are statistically different (Tukey, $p = 0.05$)

Table 5. Response of *Leucaena leucocephala* seedlings to 30 days of weathered oil exposure.

Species	Concentration	Height	Root length	Mortality	Biomass				
					Leaves	Stems	Aerial	Root	Total
	$mg.kg^{-1}$	cm		(%)	gr				
Crotalaria	150	9.8a	9.8a	2.6a	1.6a	1.0a	2.6a	0.6a	3.2a
	79,457	2.9b	2.8b	28.6a	1.0a	0.5b	1.5b	0.4b	1.9b

Values with different letter are statistically different (Tukey, p = 0.05)

Figure 4. Toxicity on *Leucaena leucocephala* seedlings with: a) 150 mg.kg^{-1} y b) 79,457 mg.kg^{-1} TPH weathered, and plants with c) 150 mg.kg^{-1} y d) 79,457 mg.kg^{-1} TPH weathered.

Some authors mention that soil conditions have a marked effect on rhizobia, because they can impact the survival and the infectivity of root hairs [68, 69]. However, there are many other factors that influence on effectiveness symbiosis such as specificity and virulence of the bacterium *Rhizobium*, nutrimental factors, soil temperature and pH [70], the latter is of utmost importance because has been reported to decrease significantly in contaminated soil [38]. Added to this there are other factors such as the accumulation of heavy metals and salts that affect soil microbiota [11, 4].

L. leucocephala almost doubled FLNFB populations with 79,457 mg.kg^{-1} of weathered oil. This can be explained because in tolerant species, some microorganisms can increase their populations in the presence of hydrocarbons [71], allowing support microbial growth. As well as [72] argue that tolerant plants are promising tools to accelerate the removal of PAH in long term polluted soils, due to their ability to thrive in a contaminated site, and its success is probably influenced by the relative amount of exudates and other compounds within the root, that stimulate microbial growth [73]. On the other hand, several authors argue that changing the C: N: P, the water content and the water retention capacity of clay soils are crucial to obtain the highest rate of degradation of TPH [74, 75].

Table 6. Response of *Crotalaria incana* to 150 days of weathered oil exposure.

Species	Concentration	Height	Root length	Knot	Leaves	Fruits	Biomass						
							Aerial	Root	Leaves	Stems	Inflorescences	Seeds	Pods
	mg.kg^{-1}	cm		Number			gr						
Crotalaria incana	150	91.7a	17.3a	16.0a	112.7a	54.0a	47.5a	5.6a	11.0a	14.7a	1.5a	9.6a	10.8a
	79,457	32.3b	7.0b	15.0a	2.7b	0.0b	0.6b	0.2b	0.1b	0.6b	0.0b	0.0b	0.0b

Values with different letter are statistically different (Tukey, p = 0.05)

Table 7. Response of *Leucaena leucocephala* to 240 days of weathered oil exposure.

Species	Concentration	Height	Root length	Knot	Leaves	Fruits	Biomass						
							Aerial	Root	Leaves	Stems	Inflorescences	Seeds	Pods
	$mg.kg^{-1}$	cm		Number			gr						
Leucaena leucocephala	150	149.0a	107.7a	43.7a	19.1a	--	88.6a	34.0a	19.1a	69.5a	--	--	--
	79,457	66.7b	7.3b	31.0b	4.6b	--	15.9b	6.9b	4.6b	11.3b	--	--	--

Values with different letter are statistically different (Tukey, $p = 0.05$)

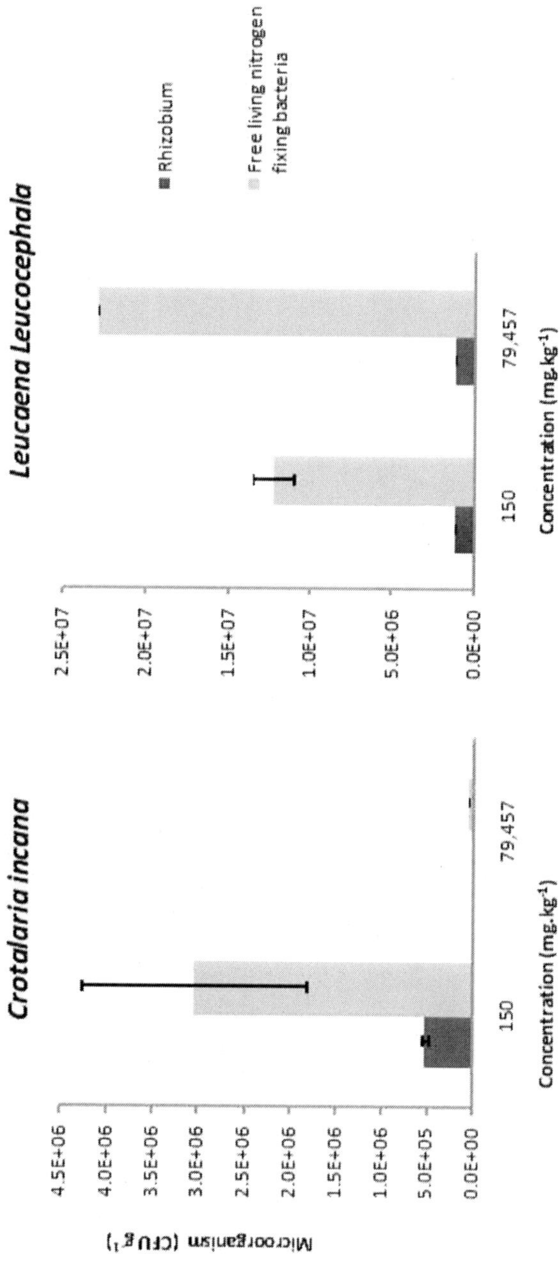

Figure 5. Quantification of populations of *Rhizobium* extracts in nodules and Free Living Nitrogen Fixing Bacteria in rhizospheric soil.

4. CONCLUSION

Soil is one of the most valuable resources that humanity has due to the variety of services offered and which depend for food. Currently, there is tremendous competition for land use, either for use as human settlements, commercial, industrial or produce raw materials (wood) and food, so weathered oil pollution is one of the most important challenges for scientists, because not only it is decontaminated, it comes to restoring the quality and safety of the soil, its properties and microbial component vital for production and proper plant development.

ACKNOWLEDGEMENTS

The author wants to especially thank Dr. Macario Vázquez Rivera, Irma Luna Capetillo, Mayra Vázquez Luna and Daniel Lara Rodríguez for their valuable support and the Technical Council of COLPOS for their scientific consultancy. Finally, the author wants to thank the Universidad Veracruzana for financing the publication of the present article.

REFERENCES

1. García-Cuellar JA., Arreguín-Sánchez F., Vázquez SH., Lluch-Cota D. B. Impacto ecológico de la industria petrolera en la sonda de Campeche, México, tras tres décadas de actividad: una revisión. Interciencia 2004 29: 311–319.

2. García-López E., Zavala-Cruz J., Palma-López DJ. Caracterización de las comunidades vegetales en un área afectada por derrames de hidrocarburos. Terra Latinoamericana 2006 (24), 17-26.

3. Olguín EJ., Hernández ME., Sánchez-Galván G. Contaminación de manglares por hidrocarburos y estrategias de biorremediación, fitorremediación y restauración. Rev. Int. Contam. Ambient. 2007 (23), 139-154.

4. Vázquez-Luna D. Environmental bases on the exploitation of crude oil in Mexico. In: Younes M. Crude Oil / Book 2. Rijeka: In Tech; 2012 p89-134.

5. Hall C., Tharakan P., Hallock J., Cleveland C., Jefferson M. Hydrocarbons and the evolution of human culture. Nature 2003 (426), 318-322.

6. PEMEX. PEMEX, 2011. Informe de responsabilidad social. D.F.: PEMEX; 2012

7. PEMEX. Petróleos Mexicanos. Informe anual 2012. D.F.: PEMEX; 2013

8. Ortínez BO., Ize LI., Gavilán, G. A. La restauración de los suelos contaminados con hidrocarburos en México. Gaceta ecológica 2003 (69), 83-92.

9. SEMARNAT-SS. NOM-138-SEMARNAT/SS-2003. Norma Oficial Mexicana que establece los límites máximos permisibles de hidrocarburos en suelos y las especificaciones para su caracterización y remediación. México: Diario Oficial de la Federación; 2003

10. Rivera-Cruz MC., Trujillo-Narcia A. Estudio de toxicidad vegetal en suelos contaminados con petróleos nuevo e intemperizado. Interciencia 2004 (29), 369-376.

11. Adams RH., Zavala-Cruz J., Morales-García F. Concentración residual de hidrocarburos en el suelo del trópico II: Afectación a la fertilidad y su recuperación. Interciencia 2008 (33), 483-489.

12. Rodríguez SM., Pereira ME., Ferreira DA., Silva E., Hursthouse AS., Duarte A. C. A review of regulatory decisions for environmental protection: Part I Challenges in the implementation of national soil policies. Environment International 2009 (35), 202-213.

13. Hernández-Acosta E., Rubiños-Panta JE., Alvarado-López J. Restauración de los suelos contaminados con hidrocarburos: Conceptos básicos. Montecillo: Colegio de Postgraduados, 2004

14. Barakat A. O., Qian Y., Kim M., Kennicutt, MC. Chemical characterization of naturally weathered oil residues in arid terrestrial environment in Al-Alamein, Egypt. Environment International, 2001 27, 291-310.

15. Kostecki P., Morrison R., Dragun J. Hydrocarbons. In Hillel D. (ed) Encyclopedia of Soils in the Environment. Oxford: Elsevier; 2005. p217- 226

16. Eweis J., Ergas S., Chang D., Schroeder E. Bioremediation principles. Series in Water Resources and Environmental Engineering. Boston: McGraw Hill; 1998

17. Riser-Roberts E. Remediation of petroleum contaminated soils. Biological, Physical and Chemical Processes. Boca Raton: Lewis Publishers; 1998

18. Botello, V. A. (1996). Características, composición y propiedades fisicoquímicas del petróleo. In: Botello VA., Rojas G., Benítez JA., Zárate-Lomelí D. (eds). Golfo de México. Contaminación e impacto ambiental: diagnóstico y tendencias. EPOMEX; 1996 400-666.

19. King RF., Royle A., Putwain PD., Dickinson NM. Changing contaminant mobility in a dredged canal sediment during a three-year phytoremediation trial. Environmental Pollution 2006 (143), 318-326.

20. González-Mille DJ., Ilizaliturri-Hernández CA., Espinosa-Reyes G., Costilla-Salazar R., Díaz-Barriga F., Ize-Lema I., Mejía-Saavedra J. Exposure to persistent organic pollutants (POPs) and DNA damage as an indicator of environmental stress in fish of different feeding habits of Coatzacoalcos, Veracruz, Mexico. Ecotoxicology 2010 (19), 1238-1248.

21. Chichizola, C. Disruptores Endocrinos. Efectos en la Reproducción. Revista Argentina de Endocrinología y Metabolismo 2003 3 (40), 172-188.

22. EA. Principles for Evaluating the Human Health Risks from Petroleum Hydrocarbons in Soils: A Consultation Paper. USA:EPA; 2003

23. López E., Schuhmacher M., Domingo JL. Human health risks of petroleum-contaminated groundwater. Env Sci Pollut Res 2008 3 (15), 278-288.

24. Totsche KU., Danzer J., Kogel-Knabner I. Dissolved organic matter-enhanced retention of polycyclic aromatic hydrocarbons soil miscible displacement experiment. J. Environ. Qual. 1997 (26), 1090-1100.

25. Charman PE., Murphy B. W. Soils their properties and management. Hong Kong: Oxford University Press; 2007.

26. Srogi K. Monitoring of environmental exposure to polycyclic aromatic hydrocarbons: a review. Environ Chem Lett 2007 (5), 169–195.

27. Gutiérrez MC., Zavala CJ. Rasgos hidromórficos de suelos tropicales contaminados con hidrocarburos. Terra Latinoamericana 2002 (20), 101-111.

28. Adetutu E M., Ball AS., Weber J., Aleer S., Dandie CE., Juhasz, AL. Impact of bacterial and fungal processes on 14C-hexadecane mineralisation in weathered hydrocarbon contaminated soil. Science of the Total Environment 2012, 414, 585-591.

29. Adetutu E., Weber J., Aleer S., Dandie CE., Aburto-Medina A., Ball AS., Juhasz, AL. (2013). Assessing impediments to hydrocarbon biodegradation in weathered contaminated soils. Journal of Hazardous Materials 2013, http://dx.doi.org/10.1016/j.jhazmat.2013.01.052 (accessed 12 June 2013).

30. Chaillan F., Chaıineau C., Point V., Saliot A., Oudot, J. Factors inhibiting bioremediation of soil contaminated with weathered oils and drill cuttings. Environmental Pollution 2006 (144), 255-265.

31. Parrish ZD., White JC., IM., Gent MP., Iannucci-Berger W., Eitzer BD., Kelsey JW., Mattina MI. Accumulation of weathered polycyclic aromatic hydrocarbons (PAHs) by plant and earthworm species. Chemosphere 2006 (64), 609–618.

32. Hernández-Acosta E., Gutiérrez-Castorena MC., Rubiños-Plata JE., Alvarado-López J. Caracterización del suelo y plantas de un sitio contaminado con hidrocarburos. Terra Latinoamericana 2006 (24), 463-470.

33. SEMARNAT. NOM-021-SEMARNAT-2000. Norma Oficial Mexicana que establece las especificaciones de fertilidad, salinidad y clasificación de suelos, estudio, muestreo y análisis. México: Diario Oficial de la Federación; 2000

34. Rivera-Cruz MC., Trujillo-Narcía A., Miranda CM., Maldonado CE. Evaluación toxicológica de suelos contaminados con petróleos nuevo e intemperizado mediante ensayos con leguminosas. Interciencia 2005 (30), 326-331.

35. Vázquez-Luna D., Castelán-Estrada M., Rivera-Cruz MC., Ortiz-Ceballos AI., Izquierdo RF. Crotalaria incana L. y Leucaena leucocephala Lam. (Leguminosae): Especies indicadoras de toxicidad por hidrocarburos del petróleo en suelo. Revista Internacional del Medio Ambiente 2010 (26), 183-191.

36. Leishman M., Wright I., Moles A., Westoby M. The evolutionary ecology of seed size. In: Fenner M. (ed) Seeds: the ecology of regeneration in plant communities. Wallingford: CAB International; 2000 p31-57.

37. Carreras M., Pascualides A., Planchuelo A. Comportamiento germinativo de las semillas de Crotalaria incana L. (Leguminosae) en relación a la permeabilidad de la cubierta seminal. Agriscientia 2001 (18), 45-50.

38. Antoniolli Z., Bellé R., Giracca E.,Thomas, P. Dormancy break of Crotalaria seeds. Ciência Rural 1993 (23), 165-168.

39. Huachin C., Carvajal J. Análisis de la semilla de doce leguminosas forrajeras tropicales. Memorias. XI Reunión Científica Tecnológica Forestal y Agropecuaria: INIFAP; 1998

40. Bosco, O. Influência de tratamentos pré-germinativos, temperatura e luminosidade na germinação de sementes de Leucena (Leucaena leucocephala (Lam.) de Wit.), cv. Cunningham. Caatinga 2009 (22), 132-138.

41. Madigan M., Martinko J., Parker J. Brock Biología de los microorganismos. Madrid, Prentice Hall Iberia; 1999

42. CIAT. 1Simbiosis leguminosa-rizobio; manual de métodos de evaluación, selección y manejo agronómico. Cali: CIAT; 1988

43. Palma-López D. J., Cisneros D. J., Moreno C. E., Rincón-Ramírez, JA. Suelos de Tabasco: su uso y manejo sustentable. Villahermosa: Instituto del Trópico Húmedo, Colegio de Postgraduados & Fundación Produce Tabasco A. C.; 2007

44. Pardo CJ., Perdomo RC., López de Mesa JL. Efecto de la adición de fertilizantes inorgánicos en la degradación de hidrocarburos en suelos contaminados con petróleo. Nova 2004 2 (2), 40-49.

45. Zavala-Cruz J., Gavi-Reyes F., Adams-Schroeder RH., Ferrera-Cerrato R., Palma-López DJ., Vaquera-Huerta H., Domínguez-Ezquivel JM. Derrames de petróleo en suelos y adaptación de pastos tropicales en el Activo Cinco Presidentes, Tabasco, México. Terra Latinoamericana 2005 (23), 293-302.

46. Hernández NL. del C. Estudio de la contaminación con petróleo de los suelos agrícolas de la cuenca baja del río Tonalá mediante espectroscopía de infrarojo cercano. Tesis de Maestría en Ciencias, Programa en Producción Agroalimentaria en el Trópico. Colegio de Postgraduados Cárdenas; 2010

47. Martínez EM., López FS. Efecto de hidrocarburos en las propiedades físicas y químicas de suelo arcilloso. Terra 2001 19, 9-17.

48. De la Garza ER., Ortiz YP., Macias BA., García C., Coll D. Actividad biótica del suelo y la contaminación por hidrocarburos. Revista Latinoamericana de Recursos Naturales 2008 (4), 49-54.

49. Cole, G. Assessment and remediation of petroleum contaminated sites. Boca Raton: Lewis Publisher; 1994

50. Fine P., Graber E., Yaron B. Soil interaction with petroleum hydrocarbons: Abiotic processes. Soil Tech 1997 (10), 133-153.

51. Romaniuk R., Brandt JF., Ríos PR., Giuffré L. Atenuación natural y remediación inducida en suelos contaminados con hidrocarburos. Ci. Suelo 2007 (25), 139-149.

52. Quiñones AE., Ferrera-Cerrato R., Gavi RF., Fernández LL., Rodríguez VR., Alarcón A. Emergencia y crecimiento de maíz en un suelo contaminado con petróleo crudo. Agrociencia 2003 37, 585-594.

53. Banks M., Schultz, K. Comparison of plants for germination toxicity tests in petroleum-contaminated soils. Water, Air, & Soil Pollution, 2005 167 (1-4), 211-219.

54. García, R. D. E. Estudio de la toxicidad de los petróleos nuevo e intemperizado en el cultivo de arroz (Oryza sativa L.). Tesis de Ingeniería Ambiental. Universidad Juárez Autónoma de Tabasco Villaermosa; 2005

55. Plice MJ. Some effects of crude petroleum on soil fertility. Soil Sci. Soc. Am. Proc. 1948 (13), 413-416.

56. Reynoso-Cuevas L., Gallegos-Martínez M., Cruz-Sosa F., Gutiérrez-Rojas M. In vitro evaluation of germination and growth of five plant species on medium supplemented with hydrocarbons associated with contaminated soils. Bioresource Technol 2008 99, 6379-6385.

57. Kisic I., Mesic S., Basic F., Brkic V., Mesic M., Durn G., Zgorelec Z., Bertovic L. The effect of drilling fluids and crude oil on some chemical characteristics of soil and crops. Geoderma 2009 (149), 209-216.

58. Tang J., Wang M., Wang F., Sun Q., Zhou Q. Eco-toxicity of petroleum hydrocarbon contaminated soil. Journal of Environmental Sciences 2011 5 (23), 845-851.

59. Salisbury F., Ross C. Fisiología de las plantas 3: desarrollo de las plantas y fisiología ambiental. Madrid: Thomson Paraninfo; 2000

60. Rao NC., Afzal M., Malallah G., Kurian M., Gulshan S. Hydrocarbon uptake by roots of Vicia faba (Fabaceae). Environ Monit Assess 2007 (132), 439-443.

61. Labud V., Garcia C., Hernández T. Effect of hydrocarbon pollution on the microbial properties of a sandy and a clay soil. Chemosphere 2007 (66), 1863-1871.

62. Rivera-Cruz MC. Clasificación de suelos tropicales influenciados por derrames de petróleo en Tabasco. Tecnociencia Universitaria 2004 (7), 6-25.

63. Li X., Feng Y., Sawatsky N. Importance of soil-water relations in assessing the endpoint of bioremediated soils. Plant Soil 1997 (192), 219-226.

64. Issoufi I., Rhykerd R., Smiciklas K. Seedling growth of agronomic crops in crude oil contaminated soil. J. Agron. Crop Sci. 2006 (192), 310-317.

65. Inckot RC., Santos GD., De Souza LA., Bona C. Germination and development of Mimosa pilulifera in petroleum-contaminated soil and bioremediated soil. Flora - Morphology, Distribution, Functional Ecology of Plants 2011 3 (206), 261-266.

66. Freedman, B. Enviromental Ecology. The Ecological effects of pollution, disturbance and other stresses. San Diego: Academic Press; 1995

67. Rivera-Cruz MC., Ferrera-Cerrato R., Volke-Haller V., Fernández-Linares L., Rodríguez-Vázquez R. Poblaciones microbianas en perfiles de suelos afectados por hidrocarburos del petróleo en el estado de Tabasco, México. Agrociencia 2002 (36), 149-160.

68. Alexander, M. Biodegradation and bioremediation. San Diego: Academic Press; 1999.

69. Langer I., Syafruddin S., Steinkellner S., Puschenreiter M., Wenzel WW. Plant growth and root morphology of Phaseolus vulgaris L. grown in a split-root system is affected by heterogeneity of crude oil pollution and mycorrhizal colonization. Plant Soil 2010 (332), 339-355.

70. Hubbell D. Legume infection by Rhizobium: a conceptual approach. Bioscience 1988 (31), 832-837.

71. Salanitro JP. Bioremediation of petroleum hydrocarbons in soil. Advances in Agronomy 2000 (72), 53-105.

72. Liste HH., Prutz I. Plant performance, dioxygenase-expressing rhizosphere bacteria, and biodegradation of weathered hydrocarbons in contaminated soil. Chemosphere 2006 (62), 1411–1420.

73. Phillips LA., Greer CW., Farrell RE., Germida JJ. Plant root exudates impact the hydrocarbon degradation potential of a weathered-hydrocarbon contaminated soil. Applied Soil Ecology 2012 (52), 56– 64.

74. Roldán-Martín A., Calva-Calva G., Rojas-Avelizapa N., Díaz-Cervantes MD., Rodríguez-Vázquez R. Solid culture amended with small amounts of raw coffee beans for the removal of petroleum hydrocarbon from weathered contaminated soil. International Biodeterioration & Biodegradation 2007 (60), 35–39.

75. Ramírez M., Zapién B., Zegarra H., Rojas N., Fernández L. Assessment of hydrocarbon biodegradability in clayed and weathered polluted soils. International Biodeterioration & Biodegradation 2009 (63), 347-353.

Chapter 6

Bioindicators and Biomarkers in the Assessment of Soil Toxicity

Carmem Silvia Fontanetti[1], Larissa Rosa Nogarol[1], Raphael Bastão de Souza[1], Danielli Giuliano Perez[1] and Guilherme Thiago Maziviero[1]

[1] Department of Biology, Institute of Biosciences, São Paulo State University (UNESP) Rio Claro, SP, Brazil

1. INTRODUCTION

Several potentially harmful chemical compounds, derived from activities of urban centres, rural properties and industries are constantly released into the terrestrial environment. In this context, the scientific community has shown increasing interest in the detection, knowledge and control of environmental agents responsible for damages to human health and sustainability of ecosystems (Magalhães & Ferrão-Filho, 2008).

Monitoring the types and quantities of toxic substances that are entering into the terrestrial environment is an exhaustive and problematic task due, mainly, to the complexity and cost resulting from the identification of the chemical substances involved. Despite the numerous analytical methods available, collecting sufficient samples in a timely fashion continues to be a great obstacle in the evaluation of environmental damages (Silva et al., 2003).

Furthermore, the determination of isolated substances by traditional chemical analyses has a limited environmental application, since it does not detect the effects on the organisms neither inform about the possible interactions between the substances (additive, antagonistic or synergistic), as well as their bioavailability (Magalhães & Ferrão-Filho, 2008). In this sense, researchers have pointed the necessity to apply biological methodologies in order to obtain an ecosystemic approach.

Biological factors may indicate better the environmental balance through the biotic indexes, derived from the observation of bioindicator species. According to Hodkinson and Jackson (2005), it is called bioindicator a species or group of species that reflects biotic and abiotic levels of contamination of an environment, presenting alterations that enables the generation of information about the quality of the environment, for example, accumulating substances in concentrations higher than those considered normal or essential for its body metabolism or presenting alterations in the number of organisms. Such organisms, due to their characteristics of little ecological tolerance to some chemical substances, can present some alteration, whether it is physiological, morphological of behavioural, when exposed to certain pollutants (Magalhães & Ferrão-Filho, 2008).

Due to their close contact with soil, some taxonomic groups of invertebrates belonging to the meso- and macro-fauna such as, Isopoda, Collembola, Oligochaeta and Diplopoda, have been proposed as bioindicator organisms (Hopkin, 2002). In addition, higher plants such as *Allium cepa* (onion), *Arabidopsis thaliana* (mustard), *Hordeum vulgare* (barley), *Tradescantia sp.*, *Vicia faba* (broad bean) and *Zea mays* (corn) are also commonly used in the assessment of soil toxicity, particularly for cytotoxicity, genotoxicity and mutagenicity assays (White & Claxton, 2004).

In a biological system, the sequential order of alterations promoted by the presence of pollutants occurs in crescent levels of biological organization, extending from the molecular or biochemical level to the physiological or individual level, until the population and ecosystem level (Stegeman et al., 1992). When a significative alteration is evident, the ecosystem must be already severely damaged. Therefore, techniques that show responses at lower levels of biological organization are considered more preventive (Nascimento et al., 2008).

Morphological alterations can be used as biomarkers in toxicity investigations of specific chemical compounds and in the monitoring of the acute and chronic effects of organisms exposed to impacted environments. In this context, the morphological analysis of target organs, carried out by ultra-morphology, histology and ultrastructure, has become widely used in studies with invertebrates, aiming to identify different damages caused by harmful substances to the organisms (Fontanetti et al., 2010).

Another tool that has shown to be increasingly efficient in the assessment of soil toxicants on the organisms is the use of molecular biomarkers. Recent studies show great interest in the use of enzymatic biomarkers as a form to monitor the environment, since the increase or inhibition of the activity of certain enzymes can explain a possible response to the environment stress.

Due to the importance to ensure the genetic integrity of the organisms, biomarkers of genotoxicity are gaining attention in the evaluation of the toxic potential of soil samples (Misik et al., 2011). The tests used in the genotoxic assessment of an agent (genotoxicity and mutagenicity tests), include the Ames test, chromosome aberrations test, micronucleus test, comet assay, SMART test (Somatic Mutation and Recombination Test), microarray and microscreen, using techniques of cellular, molecular and genetic biology both *in vitro* and *in vivo*, *in situ* and *ex situ*.

Given the above, the aim of the present chapter was to compile and discuss information present in the literature about the use of animal and plant bioindicators in the analysis of soil toxicity, as well as characterize the different biomarkers used in these organisms that enable the assessment of the soil toxicant effects in different levels of the biological scale, i.e., morphological, biochemical and genotoxic.

2. COMPLEX SUBSTANCES, ORGANIC COMPOUNDS AND METALS: POTENTIAL SOIL CONTAMINANTS

Population growth combined with the increasing industrialization is responsible for generating tons of waste per day, which, many times, are accumulated in the environment without any previous treatment. Soil becomes a cheaper and practice alternative for the final disposal of these residues, but not without consequences. Soil contamination is a broad problem, since the contaminants can be leached into groundwater, rivers and lakes. The major xenobiotics responsible for the contamination of this compartment as well as their implications for invertebrates and plants will be discussed.

2.1. Vinasse

Among the substances released into the soil with toxic potential it can be cited the vinasse, a product of the alcohol production, composed by water (97%) and a solid fraction (organic matter and mineral elements). According to Sahai et al.

(1985), due to the fast growth of the distilleries, there was, consequently, an increase in the amount of this residue, which was traditionally discharged in open areas or next to water courses, causing air, water and soil pollution. For Junior et al. (2008), the reuse of previously treated vinasse as fertilizer or soil conditioner becomes an alternative of great interest. However, Lyra et al. (2003) affirm that there are only few studies that evaluate its pollutant potential on soil and groundwater.

Thus, studies about the effects of vinasse application in the soil have been developed by researchers of different countries. Recently, Brazilian researchers have studied the genotoxicity and mutagenicity of vinasse applied *in natura* in the soil or associated with other compounds. Souza et al. (2009) used *A. cepa* to evaluate the mitotic and chromosome abnormalities resulted from exposure to landfarming soil treated with vinasse, used as a possible bioremediator. According to the authors, the vinasse was responsible for potentiating the clastogenicity of the landfarming by decreasing the pH and, thus, making available the metals that were strongly adsorbed in the organic matter of the soil.

Other studies on the vinasse toxicity were conducted by Christofoletti and Fontanetti (2010) andPedro-Escher and Fontanetti (2010) also using *A. cepa* as test organism. In the first study, the preliminary results show that the vinasse did not present cytotoxicity nor mutagenicity, but it presented genotoxic potential when applied *in natura* or associated with sewage sludge samples; the second study showed that different concentrations of vinasse diluted in water (12.5%, 25% and 50%) presented genotoxic potential and only the raw vinasse presented mutagenic potential, thus suggesting that the vinasse can cause damages in the genetic material of certain organisms.

2.2. Sewage Sludge

Another residue with pollutant potencial and problems in final disposal is the sewage sludge generated in the STSs (Sewage Treatment Stations). Its great production, mainly in the large urban centres, has led researchers to intensify the studies about the use of these waste with agricultural purposes. Therefore, recycling, via agriculture use, presents itself as a global trend (Lopes et al., 2005).

Nevertheless, sewage sludge can present, in its composition, undesirable chemicals (metals and organic chemical compounds) and biological elements (pathogens) that, in contact with man and/or fauna and flora, may cause contamination and diseases. Thus, any decision on the most appropriate final destination depends on the evaluation and minimization of the contamination risks of the environment and man (Rocha & Shirota, 1999).

Studies involving millipeds exposed to sewage sludge have shown that its components can affect the integrity of organs such as the midgut of these animals (Godoy & Fontanetti, 2010; Nogarol & Fontanetti, 2010, 2011; Perez & Fontanetti, 2011a). Mazzeo et al. (2010), using *A. cepa,* investigated the genotoxic and mutagenic potential of domestic sewage sludge at different concentrations.

2.3. Polycyclic Aromatic Hydrocarbons (Phas)

Some of the main pollutants that cause concern in relation to soil contamination are the PAHs (Bispo et al., 1999), which are compounds formed by two or more benzene rings, exclusively constituted by atoms of carbon and hydrogen (Netto et al., 2000). There are many origin sources but it can be highlighted industrial processes, such as petroleum refining, combustion of organic matter and burning of coal (Page et al., 1999). According to the IUPAC (International Union of Pure and Applied Chemistry) there are, currently, over 100 known PAHs, however only 16 have environmental and toxicological importance (Environment Protection Agency [EPA], 1986). These compounds are able to react, directly or after undergoing metabolic transformations, with the DNA, becoming potential carcinogens and efficient mutagens.

Soil receives considerable amounts of PAHs that, because of the complexity of their chemical structure, low solubility in water and strong sorption tendency into the soil, become recalcitrant and remain for long periods in the environment, enhancing the probability of exposure of humans and animals to these compounds (Jacques et al., 2007). When present in the environment they can be transferred to invertebrates by ingestion of soil and plant material contaminated or by cuticle (Achazi & Van Gestel, 2003).

In order to reduce the negative impact of these compounds in the soil, petroleum refineries use a bioremediation system called landfarming. The technique has been used for the treatment of soils contaminated with hydrocarbons for 100 years and, the petroleum industry for at least 25 years (Riser-Roberts, 1998). Currently, other types of industry started to employ this technique, such as textile and food industries and treatment of effluents.

Souza et al. (2009) carried out bioassays with *A. cepa* in order to assess landfarming soil samples before and after biodegradation of hydrocarbons.

Before biodegradation, the landfarming had 13.5 g/Kg of Total Petroleum Hydrocarbons (TPH) and caused strong clastogenic and mutagenic effects. After 108 days of biodegradation, the concentration of TPH decreased 27% with significant reduce of mitotic and chromosome abnormalities, micronuclei and nuclear buds.

Using the diplopod *R. padbergi* exposed to different concentrations of industrial soil contaminated with PAHs, Souza et al. (2011) and Souza and Fontanetti (2011) analyzed the perivisceral fat body and midtgut of the animals and verified that there were several alterations in these two tissues.

2.4. Dioxins

Seven dibenzo-p-dioxins (PCDDs), 10 polychlorinated dibenzofurans (PCDFs) and 12 polychlorinated biphenyls (PCBs) are called dioxins (Word Health Organizatrion [WHO], 2010; United States Environmental Protection Agency [USEPA], 2000, 2003), being released into the environment as a byproduct of chemical processes and through the combustion of industrial and municipal wastes (Stephens et al., 1995). According to Schlatter (1994), since the accident in Seveso, Italy, they became the symbol of threat caused by toxic chemicals. As a result of widespread fear, dioxins are a matter of real concern in relation to environmental contamination.

Among the isomers of PCDDs, the most toxic is 2,3,7,8 - tetrachloro-dibenzo-para-dioxin (2,3,7,8-TCDD) (Eisler, 1986). There is little information in literature on the effects of PCDDs on terrestrial invertebrates. Reinecke and Nash (1984) reported that two species of earthworms (*Allolobophophora caliginosa* and *Lumbricus rubellus*) showed no adverse effects when exposed for 85 days in soil with 5 ppm of 2,3,7,8-TCDD, but both species died at 10 ppm. Studies involving plants and PCBs have been made, since these organisms are less sensitive to PCBs and thus may be a possible route of biomagnification in various food chains (Sinkkonen et al., 1995).

2.5. Agrochemicals

The use of fertilizers and pesticides has become a common practice due to population growth, food crisis and consequent need for the increase in the agriculture production. Within the existing agriculture model, agrochemicals are classified as one of the main chemical pollutants that are disseminated throughout the planet (Grisolia, 2005). However, there is still little information about the effects of these chemical compounds on invertebrates that occupy levels of high sensitivity in trophic chain (Mantecca et al., 2006).

Among the studies carried out with terrestrial invertebrates it was analyzed the possible alterations in the biomass (Niemeyer et al., 2006 a), reproduction (Helling et al., 2000), behaviour (Niemeyer et al., 2006b), survival (Diao et al., 2007) and tissular and cellular lesions (Nasiruddin & Mordue, 1993) resulted from exposure to certain agrochemicals. Associated with the use of bioindicators and biomarkers it is also used the direct analysis of the presence of residues in soil samples by specific equipments such as the spectrophotometer or chromatograph.

Pesticides have been widely tested by bioassays with plants and positive results are usually obtained (Leme & Marin-Morales, 2009). A clear example is the *A. cepa* test successfully used in the evaluation of the mutagenic and genotoxic potential of herbicides such as trifluralin (Fernandes et al., 2007;2009).

2.6. Metals

Heavy metals or trace metals are terms applied for a great amount of trace elements that are industrially and biologically important. From the point of view of human health, agriculture and ecotoxicology, the most worrying heavy metals are As, Cd, Hg, Pb, Ti and U. Studies involving heavy metals in ecosystems have shown that many areas near urban centres, mines and road systems have high concentrations of these elements (Alloway, 1994). Metals are highly persistent in the soil with persistence of up to thousands of years (Mc Grath, 1987) and can express their pollutant potential directly on the soil organisms by availability to plants and transference to the food chain, both by plants and by the contamination of superficial waters or groundwater (Chang et al., 1987).

The main anthropogenic sources of metals are fertilizers, pesticides, contaminated irrigation water, combustion of coal and oil, vehicular emissions, incineration of urban and industrial wastes and, mainly, mining and smelting (Tavares & Carvalho, 1992).

Due to their habits in the superficial layers of the soil, invertebrates of the saprophagous fauna, such as isopods, diplopods and springtails are regularly exposed to metals (Hopkin, 2002). Heikens et al. (2001) carried out a literature review to clarify the concentration of metals in terrestrial invertebrates and they concluded that the concentration in most of the groups happened in the order Pb>Cd>Cu. Afterwards, Köhler (2002) conducted a study to determine the location of these metals in the bodies of soil arthropods. The genotoxic potential of metals has also been studied by several authors using plants as test systems (Knasmüller et al., 1998; Rank & Nielsen, 1998).

3. INVERTEBRATES OF THE EDAPHIC FAUNA AND HIGHER PLANTS AS SOIL BIOINDICATORS

One important question in ecotoxicological studies refers to the choice of the bioindicator species. It will depend on its ecological and toxicological importance, facility to be maintained in laboratory, reproductive rate and sensitivity (it must be affected by several chemical agents but less affected by abiotic factors) (Römbke & Garcia, 2000). Many authors agree that the main features needed to be a good bioindicator are sensitivity, good representativeness and functional importance in the ecosystem, as well as easy collection, identification and analysis (Greensdale, 2007). In this context, some taxonomic groups of soil invertebrates and higher plants have been proposed as bioindicator organisms.

3.1. Terrestrial Invertebrates

Terrestrial arthropods of the saprophagous fauna such as, Isopoda, Collembola and Diplopoda are among the most appropriate organisms to evaluate the effects of the accumulation of toxic substances present in the soil, due to their direct contact with contaminants present in it (Gräff et al., 1997;Hopkin et al., 1989). Annelids, in special the Oligochaeta, are also frequently used in toxicity tests. These invertebrates get in contact with a great variety of pollutants present in this compartment by their movement and ingestion of contaminated soil or leaf litter (Spadotto et al. 2004).

Oligochaeta are considered one of most important representatives of the edaphic macro-fauna (Kale, 1998). Several factors make earthworms excellent bioindicators of the toxicity of chemical substances in the soil, such as the knowledge already accumulated on their habitats and important trophic position of these invertebrates, which are situated in the lowest levels of the terrestrial food webs, serving as food for several animals and route of transference and biomagnification of contaminants along these webs (Andréa, 2010).

Due to their great importance in the soil, their wide distribution and all the reasons previously cited, earthworms, mainly the species *Eisenia fetida* (figure 1) and *E. andrei* were chosen for several toxicity tests for registration of agrochemicals in the regulatory agencies of several countries, including Brazil (Andréa, 2010). Other species such as *Lumbricus terrestris* and *L. rubellus* have been widely used in studies of bioaccumulation of metals (Amaral & Rodrigues, 2005; Veltman et al., 2007).

Figure 1. Earthworm *Eisenia fetida.* (Photo: Raphael Bastão de Souza and Larissa Rosa Nogarol)

Collembola are among the most important members of the soil meso-fauna involved in the decomposition process and are vulnerable to the effects of its contamination (Bengtsson & Rundgren, 1984). Greensdale (2007) lists some favourable points in choosing Collembola as bioindicators, such as presence in all ecosystems, abundance and ease of collection in sufficient number to allow statistical analyses. Moreover, they have short life cycle, making that they respond quickly to environmental changes and, as they are in direct contact with the soil, they are more sensitive to some type of stress applied in the ecosystem.

Several studies point out this organism as bioindicator, applying different methodologies and evaluation parameter. Tests of reproduction associated to survival rates (Pedersen et al., 2009; Sverdrup et al., 2010;) and evaluation of abundance and/or diversity of species in areas that suffer some type of degradation (Sousa et al., 2004) are the most used methodologies.

Another taxonomic group used in toxicological analyses is Isopoda, one of the largest orders of crustaceans with approximately 10,000 thousand described species, mostly marine (Schultz, 1982). Terrestrial isopods have already been used in toxicity tests of soil and the main parameters of evaluation were abundance of individuals (Faulkner & Lochmiller, 2000), reproduction rates (Niemeyer et al., 2009) and survival (Stanek et al., 2006).

Metals are the main toxic agents evaluated using Isopoda, since these invertebrates bioaccumulate these elements. In this sense, researchers have carried out studies on bioaccumulation (Blanusa et al., 2002; Hopkin et al., 1993), the cytotoxic effect of metals (Köhler et al., 1996a; Odendaal & Reinecke, 2003) and the detoxification mechanisms (Hopkin, 1990; Köhler & Triebskorn, 1998) and the terrestrial isopod *Porcellio scaber* is the most studied.

The importance of diplopods in the recycling of nutrients, aeration and fertilization of soil is frequently mentioned in the literature (Dangerfield & Telford, 1989). Due to the habits of the diplopods, colonizers of various soil layers, these animals can be greatly influenced by the deposition of metals, organic compounds and complex substances in the soil.

Most studies in the literature using diplopods as bioindicators of the soil are related to metals. However, the effect of organic pollutants and complex mixtures on these invertebrates is relatively little known (Souza & Fontanetti, 2011). In this context, the first study carried out with diplopods, as possible bioindicators, was conducted by Hopkin et al. (1985), involving the assimilation of metals by the species *Glomeris marginata*. In this study, it was verified a higher uptake of copper, zinc and cadmium by the animals collected in soils contaminated when compared to those animals collected in non-contaminated environments. The authors comment that ultrastructural studies of different organs would be necessary to understand the "metals path", particularly in the gut of these invertebrates.

Later, Triebskorn et al. (1991) exposed several invertebrates such as mites, insects and diplopods to different toxic substances and used the ultrastructural analysis in order to demonstrate the applicability of using such animals in biomonitoring. In the study carried out by Köhler et al. (1992), it was analyzed the impact of lead on the efficiency of assimilation in diplpods, submitted to different environmental conditions. The researchers used different species of diplopods and found that only*Glomeris conspersa* increased the ingestion of food containing lead when compared to a non-contaminated diet.

Recently, the toxicity assessment of complex substances was performed with the Brazilian species*Rhinocricus padbergi* (figure 2) exposed to different concentrations of sewage sludge (Godoy & Fontanetti, 2010; Nogarol & Fontanetti, 2010, 2011; Perez & Fontanetti, 2011a) and landfarming (Souza & Fontanetti, 2011 ; Souza et al., 2011). The histological and histochemical analysis, as well as the ultrastructural analysis, showed that such substances are toxic to the diplopod studied, since different tissular and cellular alterations were observed in the midgut and perivisceral fat body of these invertebrates.

Figure 2. Diplopod *Rhinocricus padbergi*. (Photo: Larissa Rosa Nogarol and Raphael Bastão de Souza)

3.2. Higher Plants

Plants, despite their structure and metabolic differences, can offer important information about the cytotoxic, genotoxic and mutagenic potential of substances, even when exposed in short term and offer some advantages such as low cost cultivation and easy maintenance, comparatively, to mammals (Rodrigues et al., 1997). In studies with complex mixtures, plants have also shown satisfactory results, indicating that plants are sensitive enough to detect the adverse effects of environmental samples (Majer et al., 2005).

Plants can be directly exposed to the contaminant, without any dilution or filtration of the sample (Steinkellner et al., 1999). Moreover, Grant (1994) cites other advantages of employing higher plants: (1) higher plants are eukaryotes, thus, their structure and cellular organization are similar to that of humans and it is possible to establish comparisons with animals; (2) the techniques employed for the study are relatively simple and can be performed with agility; (3) cultivation of the organisms has low cost and easy maintenance; (4) the assays can be carried out under a wide range of environmental conditions, pH and temperature; (5) higher plants can regenerate easily; (6) assays with higher plants can be used to assess the genotoxic potential of simple substances or even complex mixtures; (7) it can be used for *in situ* monitoring; (8) can be used for monitoring for several years and are highly reliable; (9) studies have shown correlations with cytogenetic assays in mammals; (10) can be used together with microbial assays to detect mutagenic metabolites (pro-mutagens); (11) genotoxicity studies with plants are presenting high sensitivity in tests with carcinogenic agents.

On the other hand, according to Majer et al. (2005), one of the limitations of using plants as bioindicators is the lack of sensitivity for certain classes of pro-mutagens such as the nitrosamines, heterocyclic amines and some classes of PAHs. In contrast, Ventura (2009) showed that the *A. cepa*system is susceptible to nitro aminobenzene, while Mazzeo (2009) observed the same effect for benzene, toluene, ethylbenzene and xilene (BTEX).

Among the higher plants, onion (*A. cepa*) is the most used plant to determine the cytotoxic, genotoxic and mutagenic effects of many substances present in the soil. Its cellular kinetics characteristic favours a rapid growth of the roots, due to the great number of cells in division. Therefore, the record of the mitotic activity and abnormalities in the cell cycle of the meristematic cells of its roots can be easily visualized (Grant, 1994). Leme and Marin-Morales (2009) affirm that the *A. cepa* test is a fast and sensitive technique to detect genoxotic and mutagenic substances dispersed in the environment.

The evaluation of genetic alteration can be also performed using different species of the genus*Tradescantia* (figure 3) through the detection of mutations induced by agents present in the air, soil and water by the analysis of micronuclei in the mother cell of the pollen grain (Trad-MCN). The species *Tradescantia* are specially indicated for direct application in regions and countries in development due to the advantages such as easy handling and relatively low maintenance cost (Shima et al., 1997).

Vicia faba is a popular material that has been widely used not only in cytological studies, but also in physiological experiments (Kanaya et al., 1994). This organism was initially used in radiobiological tests in investigation of mechanisms of formation of chromosomal aberrations by ionizing radiation (Read, 1959). Later, Kihlman (1975) developed and standardized the *V. faba* meristematic cell bioassay for analysis of chromosomal aberrations, and since then has been widely used for genotoxicity studies for evaluation of sister chromatid exchange (Kihlman & Kronborg, 1976; Kihlman & Andersson, 1984). This technique is very similar to *A. cepa* test; the method does not require sterile conditions or any material or equipment of high cost. Further details of this test are described by Kihlman (1975).

Figure 3. *Tradescantia pallida*. (Photo: Guilherme Thiago Maziviero)

4. BIOMARKERS

Molecular, biochemical and physiological compensatory mechanisms can become operative in organisms after exposure to environmental contaminants. This may result in the inhibition or facilitation of one or more physiological mechanisms or functional and structural changes. In this sense, the use of biomarkers allows obtaining information about the biological effects of pollutants and mechanisms of action of xenobiotics on the fauna.

Several authors have proposed different definition for the term biomarkers. According to Lam and Gray (2003), biomarkers can be defined as biochemical, cellular or molecular alterations or physiological changes in the cells, body fluids, tissues or organs of an organism that are indicative of exposure or effect of a xenobiotic. Despite being older, the definition proposed by Depledge (1993) and Depledge et al. (1993) has a more comprehensive character and it is considered the most widely used nowadays: biomarkers are defined as adaptive biological responses to stressors, evidenced as biochemical, cellular, histological, physiological or behavioural alterations.

In the scope of measuring the toxic effects in the organisms at a cellular or molecular level, biomarkers represent an initial response to environmental disturbances and contamination. Therefore, they are generally considered more sensitive than the tests that measure these effects at higher levels of biological hierarchy, such as individual or population (Mc Carthy & Shugart, 1990).

Thus, during the last decades, several biomarkers have been used effectively, especially as tests for specific toxicants, since biomarkers when combined with biomonitors can create a sophisticated multiple target system to detect a variety of environmental hazards in a fast and economically feasible way, in a single test organism, helping in the establishment of priorities for action in the control of environmental pollution.

4.1. Morphological Biomarkers

The detection of many classes of damage in several tissues and cellular types becomes possible by using morphological biomarkers. Such morphological alterations may provide qualitative evidences of a functional adaptation to the external environment (Meyers & Hendricks, 1985). Moreover, the qualitative assessment of such changes before the death of the organism may provide early indications of toxicity (Nogarol & Fontanetti, 2010; Triebskorn et al., 1999).

For such analysis, histology and ultrastructure are used. By these techniques it is possible to diagnose cellular and sub-cellular symptoms resulted from intoxication as well as locate symptoms of cellular death and reveal reactions in response to chronic and sub-lethal exposure in cells and tissues (Fontanetti et al., 2010; Kammenga et al., 2000).

Studies show that one of the main contaminants of the soil, metals, are selectively concentrated in only one or few organs, or in specific regions of the tissues in most of soil invertebrates and typically these organs are part of the digestive tract (Dallinger, 1993). For example, in millipeds (Köhler & Alberti, 1992), isopods (Dallinger & Prosi 1988) and springtails (Pawert et al., 1996), the epithelium of the midgut is the main target of metals. Thus, the epithelium of the digestive tract represents the first barrier against the intoxication of the whole organism (Walker, 1976).

In diplopods, some studies with this approach were performed using the digestive tube and the fat body (Hopkin et al., 1985; Köhler & Triebskorn, 1998; Triebskorn et al., 1999). Morphological alterations observed in the midgut (figure 4) and in the perivisceral fat body (figure 5) of the diplopod *R. padbergi* were successfully used as sublethal biomarkers in the evaluation of soils contaminated with complex substances such as sewage sludge (Godoy & Fontanetti, 2010; Nogarol & Fontanetti, 2010,2011; Perez & Fontanetti, 2011a) and landfarming (Souza & Fontanetti, 2011 ; Souza et al., 2011).

In the studies performed with the diplopod *R. padbergi*, it was possible to observe tissular and cellular responses related to detoxification mechanisms such as increased cytoplasmic granules (spherocrystals) and intense release of secretory vesicles into the intestinal lumen of these invertebrates (Nogarol & Fontanetti, 2010; Perez & Fontanetti, 2011a). These secretory vesicles of the apocrine type seems to help in the detoxification of toxic substances initially absorbed by the organism and form a protector layer that would reduce the contact between the toxic agent and the intestinal epithelium.

The formation of agglomerates of haemocytes through the cells of the "fat body" layer was also observed and this response is directly related to a defence mechanism of the animal. According to van de Braak (2002), haemocytes can migrate to the injury site in the tissue by a chemotaxis process that results in inflammation. By this inflammatory reaction, these cells act in the removal of toxins and possibly help in the re-absorption of the damaged epithelium in order to maintain the homeostasis of the organism. In a recent review conducted by Perez and Fontanetti (2011b) it becomes clear that this tissular response is common in different invertebrates exposed to environmental stress conditions. According to the authors, the monitoring of the number of haemocytes can be used as a measure of stress in sentinel species due to environmental contamination.

The mechanisms of defence and detoxification require high and continuous energy expenditure, especially when the organism is exposed to a toxic agent for a long period. In this sense, histological and ultrastructural studies showed some of the main responses of this invertebrate related to higher energetic needs. Nogarol and Fontanetti (2011) observed at an ultrastructural level a high increase in the number of tracheioles between the cells of the "fat body" layer that compose the midgut of diplopods sub-chronically exposed to sewage sludge. The authors suggest that a higher oxygenation of the tissue was necessary to enable the formation of molecules of adenosine triphosphate (ATP), used in the detoxification mechanisms.

Toxic agents may be able to cause cellular death by necrosis, evidenced mainly by the intense cytoplasmic vacuolization in the principal cells of the midgut epithelium of diplopods exposed to landfarming (Souza & Fontanetti, 2011) and sewage sludge (Nogarol & Fontanetti, 2011; Perez & Fontanetti, 2011a). In addition to the cytoplasm, other cellular compartments were affected by the exposure to toxic agents leading to cellular inviability. In these cases, damaged cells are expelled towards intestinal lumen.

Samples of landfarming and sewage sludge presented genotoxic action, evidenced by the occurrence of nucleus fragmentation in the principal epithelial cells, karyolysis in the nucleus of the cells of the fat body layer (Souza & Fontanetti, 2011) and loss of integrity of the nuclear envelope of hepatic cells and cells of the "fat body" layer of the midgut (Nogarol & Fontanetti, 2011).

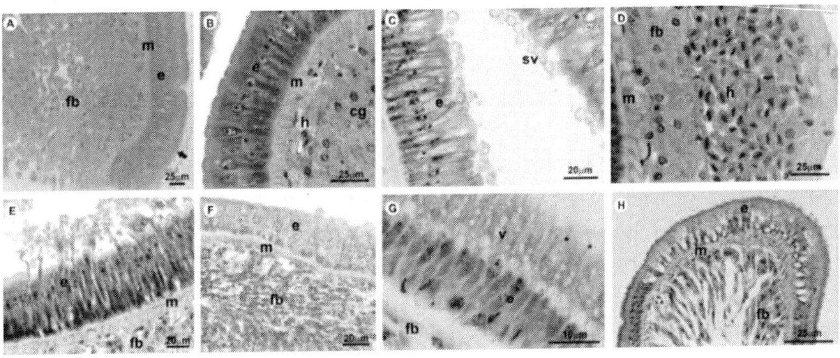

Figure 4. Midgut of the diplopod *R. padbergi* stained with Hematoxylin-Eosin. Unexposed animals (A; B); Animals exposed to sewage sludge (C-H). secretion vesicles (C); haemocytes aglomeration (D); epithelium renewal (E); increase of cytoplasmatic granules in "fat body layer" (F); cytoplasmatic vacuolization (G); volume reduction of the cells in "fat body" layer of midgut (H). e=epithelium; m= muscle layer; fb= "fat body" layer; h= haemocytes; v= vacuole; sv= secretion vesicle; * dilatation of intercellular space (Photos: Larissa Rosa Nogarol; Raphael Bastão de Souza and Tatiana da Silva Souza)

In diplopods, the fat body, both parietal and perivisceral, is constituted of trophocytes and oenocytes (Fontanetti et al., 2004) and presents intense metabolic activity, such as storage of lipids, uric acid and proteins as well as storage, neutralization and excretion of substances that are not useful (Fontanetti et al., 2006; Hopkin & Head, 1992; Hubert, 1979). In this sense, Souza et al. (2011) exposed diplopods of the species *R. padbergi* in bioassays containing industrial soil contaminated by PAHs and metals (landfarming) in order to analyze histological and histochemical alterations in the perivisceral fat body. The authors concluded that the fat body can be used as a target organ and that the alterations observed, such as loss of integrity of the plasmatic membrane, cytoplasmic disorganization and depletion of energetic reserves can be considered stress biomarkers in this animal. Similar responses were observed in animals exposed to sewage sludge (Abe et al., 2010).

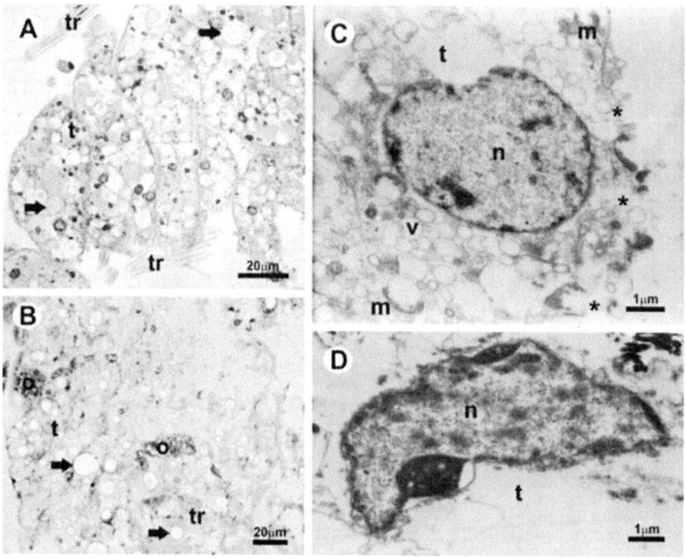

Figure 5. Perivisceral fat body of the diplopod *R. padbergi* stained with Hematoxylin-Eosin (A; B) and submitted to TEM routine (C; D). Unexposed animal (A); Animal exposed to sewage sludge (B-D). Loss of cell limit and increase of spherocrystal (B); Cytoplasmatic vacuolization and loss of cell membrane integrity (C); Nucleus deformation (D). t= trophocyte; tr = tracheoles; o= oenocyte; m= mitochondria; n= nucleus; v= vacuole; arrows= spherocrystals; *= loss of cell membrane integrity. (Photos: Raphael Bastão de Souza and Larissa Rosa Nogarol)

4.2. Genotoxicity Biomarkers

The increase in the genotoxic load in the terrestrial ecosystems by the release of chemical products and physical agents can cause impact on the organisms, inducing increase in the frequency of mutations; such effects can lead to a decrease in the size of the population and, eventually, extinction of species and consequently affect the stability of this ecosystem (Majer et al., 2005). In this sense, it became necessary to develop different tests to evaluate the genotoxic potential of soil samples.

Due to the highly conserved structure of the genetic material, it is possible to use a wide variety of species in genotoxicity tests; currently, the most widespread methods for the routine tests are based on the use of indicator bacteria and also basidiomycetes fungi, plants, insects and cultured mammalian cells or even laboratory animals for mutagenicity tests.

According to the literature, the Ames test is the most widely used in genotoxicity evaluations of soils and leachate (Claxton et al., 2010; Wölz et al., 2011). This test, also known as *Salmonella*/microsome, consists, basically, in the employment of strains of the auxotrophic bacteria *Salmonella typhimurium*, i.e., deficient in the synthesis of the aminoacid histidine; the strains of these cells are unable to grow in minimum medium, where the mutagenic compounds are able to restore the synthesis capacity of this aminoacid, thus, the mutagenic expression corresponds to the growth of the colony in a minimum culture medium and it can be easily detected by counting the colonies (Umbuzeiro & Vargas, 2003). However, due to the low sensitivity of the Ames test for heavy metals, more studies should be directed to the development of bioassays with higher organisms (Gatehouse et al., 1990, as cited in Lah et al., 2008).

Meristematic cells of *A. cepa* and *V. faba*, for example, constitute an effective cytogenetic material to analyze chromosome aberrations (figure 6) caused by soil pollution. The use of meristematic cells makes possible the quantification of several morphological and cytogenetic parameters (endpoints), including the morphology and growth of roots and determination of several parameters of cytotoxicity, genotoxicity and mutagenicity. The analysis of the cytotoxicity can be done by determining the mitotic index and cell death. The induction of aberrant metaphases, anaphases and telophases, such as bridges, loss and chromosome stickiness, polyploidy, irregular nuclei and nuclear buds are parameters for the genotoxicity analysis, while the micronuclei and chromosome breaks allow the mutagenicity analysis (Fernandes et al., 2007; Leme & Marin-Morales, 2008; Souza et al., 2009).

Chromosome aberration test concerns the discovery of the mechanisms of action of a particular agent, since the division process is well known. Kovalchuk et al. (1998) state that the chromosome aberrations assay with *A. cepa*, can be used as a tool for quantifying and monitoring genetic alterations in soils radioactively contaminated. Moreover, the chromosome aberration assay in onion was the first of nine plant systems accepted in the Genotoxic Program of the Environmental Protection Agency (USEPA) and is widely used for monitoring residual water. Such sensitivity is attributed by Ma et al.(1995), to the large size of the chromosomes and because they are mostly metacentric. On the other hand, bioassays based on chromosome aberrations, in certain cases, tend to be replaced by less time-consuming techniques, such as the micronucleus assay (MN), which can be performed with mitotic cells in roots (of *V. faba* or *A. cepa*) or meiotic cells, in tetrads of *Tradescantia* (Misik et al., 2011).

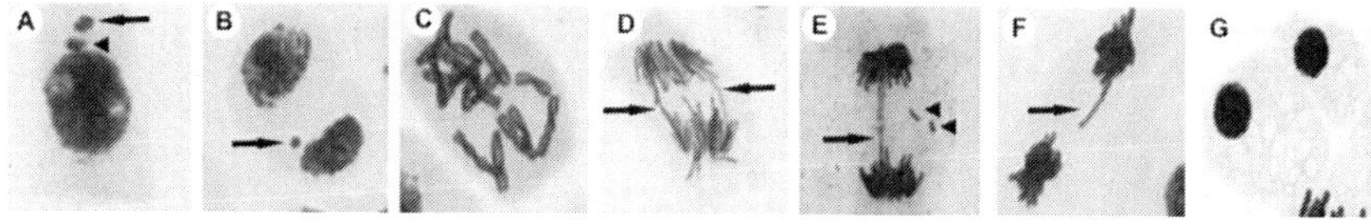

Figure 6. Alterations observed in *A. cepa* meristematic cells. (A) nuclear bud (arrow head) and micronucleus (arrow); (B) micronucleus (arrow); (C) C-metaphase; (D) anaphase with chromossomal bridge (arrows); (E) telophase with cromossomal bridge (arrow) and chromossomal break (arrows head); (F) telophase with cromossomal delay (arrow); (G) cell death. (Photos: Cintya Aparecida Christofoletti)

The micronucleus test in *Tradescantia* (Trad-MCN) (figure 7) is a sensitive mutagenicity test, of short exposure and simple evaluation, applicable in the species *T. pallida* and in the clones BNL 4430 and KU 20 (Misik et al., 2011). Besides the Trad-MCN, it is possible to evaluate mutations in somatic cells of the staminal hair (Trad-SHM) in young inflorescences of the hybrid clone BNL 4430 (Brookhaven National Laboratory). However, currently, the clone KU 20 (Kyoto University) is more applicable to this technique due to the higher number of inflorescences per cycle. The mutation results in the expression of the recessive allele, which implies in the phenotype of pink colouration. The high rate of pink cells, as well as the loss of reproductive capacity are indicative of mutagenicity (Ma et al., 1996).

4.3. MOLECULAR BIOMARKERS

The use of molecular biomarkers in the environment monitoring represents a significative tool for the evaluation of the contamination in different organisms. Despite morphological markers provide good qualitative evidence of damages caused by certain pollutants, it is known that biochemical alterations resulted from the toxic action of a contaminant are early evidence of negative effects of the exposure, since they precede the onset of visible damages. Following, it is presented some biochemical biomarkers extensively used in studies of environmental impact, including ecotoxicological analyses of the soil.

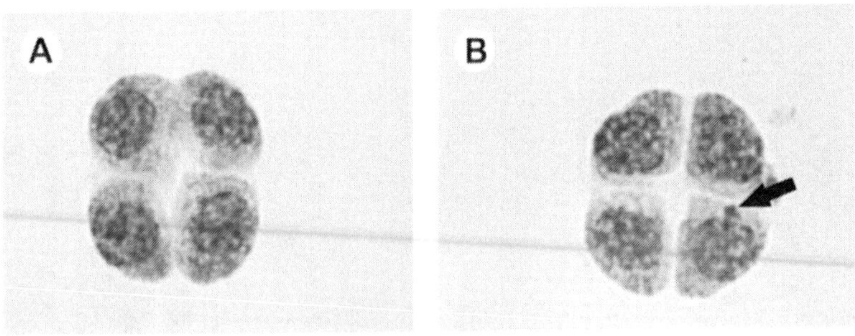

Figure 7. Micronucleus in polen cells of *Tradescantia* (arrow in B). (Photos: Janaína Pedro-Escher)

4.3.1. Enzymatic Antioxidants

The use of enzymatic activity as biomarker is due to the fact that toxic compounds have high affinity for electron pairs found in the aminoacids that form the enzymes (Cogo et al., 2009). One of the main monitored parameters in ecotoxicological analyses of the soil is the concentration of metals. Exposure to metals can intensify the production of reactive oxygen species (ROS), which are normally produced in non-stressed cells and their excess can lead to the oxidative stress and cause harmful effects (Barreiros et al., 2006).

When a cell undergoes oxidative damage, the injuries are minimized in the different organisms by enzymatic and non-enzymatic antioxidants (Freitas et al., 2008). Among the enzymatic antioxidants, some examples would be the superoxide dismutase, catalase, gluthathione reductase and gluthathione -S-transferase (Mishra et al., 2006).

Superoxide dismutase catalyzes the formation of H_2O_2 from O_2. This enzyme was the first discovered among the enzymatic antioxidants and, generally is one of the first to act against damages caused by ROS (Nordberg & Arnér, 2001).

Now, the catalase function is to facilitate the removal of H_2O_2, degrading it in H_2O and O_2. Thus, it reduces the risk of forming the radical hydroxyl from H_2O_2, since this oxygen reactive species is one of the most harmful to the biological systems (Betteridge, 2000; Diplock et al., 1998).

The metabolism of gluthathione is one of the main antioxidant defence mechanisms in the living systems (Valko et al., 2006) and specific metals can induce the synthesis of this compound in different species (Backor et al., 2007). In order to perform its function as oxidant agent, gluthathione must be in its reduced form, reaction catalyzed the enzyme gluathathione reductase (Creissen et al., 1994).

Another important defence system against the increase of free radicals involves the enzyme gluthathione peroxidase, which acts in the removal of hydrogen peroxide and lipid peroxides from the cell (Rover Junior et al., 2001). One of the forms of the gluthathione peroxidase is the gluthatione-S-transferase, one of the most studied detoxicant enzymes in different organisms, since it has an essential role in the cellular response to the stress caused by herbicides in plants. It is considered a detoxification enzyme because it metabolizes a great variety of xenobiotic compounds, catalyzing their conjugation with the reduced molecule of gluthathione and forming substances of low toxicity (Malmezat et al., 2000).

According to Almeida (2003), depending on the type of contaminant and exposure period of the organism to the contaminated environment, the activity of the antioxidant enzymes can be stimulated or inhibited. Generally, the increase of the enzyme activity results from an increase in the production of ROS, which leads to a exacerbated induction of enzymes; now, the decrease can be related to prolonged exposure of the organism to environments highly contaminated, where the production of ROS and the consequent deleterious effects of such production surpasses the defence efforts of the organism.

Several other studies describe alterations in the enzymatic activities of the superoxide dismutase, catalyse, gluthathione reductase and gluthathione peroxidise in different organisms exposed to stress conditions, especially metals, corroborating their use as effective biochemical biomarkers in the evaluation of environmental impacts (Bocchetti et al., 2008; Cogo et al., 2009).

4.3.2. Heat Shock Proteins

All organisms, from bacteria to mammals, respond to different environmental stress conditions by the synthesis of highly conserved proteins, known as heat shock proteins (HSPs) (Hamer et al., 2004). They are so called because they were first described in cells of *Drosophila melanogaster* during exposure to high temperatures. At the time, it was verified that the exposure of cells to heat produced a new pattern of thickening of chromosomes, which represented specific sites of transcription for the synthesis of proteins. The stress induced the expression of certain genes, which led the cell to produce a certain class of proteins, so-called heat shock proteins. Later, researchers observed that these proteins were expressed in almost all living beings, and not only in response to heat, but also when the cell was exposed to a series of other stressing factors (toxic concentrations of metals, organic pollutants, temperature, osmolarity, hypoxia/anoxia and ultraviolet radiation), and then they began to be called "stress proteins" or "anti-stress proteins" (Meyer & Silva, 1999).

Many toxicants affect the correct conformation and consequently, the function of different proteins. In this condition, where the proteins are found incorrectly folded inside the cell, it is initiated a stress response. The HSPs take action, acting as molecular chaperones, since they bind to other proteins, regulating their conformation, movement through the membrane or organelles and enzymatic activity (Calabrese et al., 2005). Therefore, they avoid incorrect interactions between proteins, helping in their synthesis, folding and degradation (Meyer & Silva, 1999).

According to Bierkens (2000), HSPs are one of the main cellular markers in the evaluation of the toxicity of different compounds and are widely used to monitor ecosystems. Such monitoring has shown high levels of HSPs in the tissues of invertebrates collected in contaminated areas, when compared to those animals existent in uncontaminated environments (Bierkens, 2000; Malaspina &Silva-Zacarin, 2006).

These proteins are found highly conserved in all the living organisms (Burdon, 1986) and can be classified according their molecular weight into four families: HSP90 (90 kDa), HSP70 (70 kDa), HSP60 (60 kDa) and small HSPs. The family HSP70 is one of the most studied and several studies have shown its induction in stress conditions by heavy metals (Köhler et al., 1992 , 1996b; Nadeau et al., 2001; Zanger et al., 1996).

Monari et al. (2011) worked with the mollusc species *Chamelea gallina*, observing an increase in the expression of HSP70. The authors affirm that the induction of HSP70 can be considered an adaptation mechanism associated with changes in the environmental parameters.

Silva-Zacarin et al. (2006), using immune-histochemical methods, observed an increase in the levels of the products of the positive reaction to HSP70 in the salivary glands of bees treated with acaricides in comparison with the control group. Moreover, they also verified alterations in the immune-reactivity between the nucleus and cytoplasm according to the acaricide used and the treatment period. According to the authors, the determination and location of HSP70 by immune-histochemistry can be useful to detect cellular responses to chemical stressors.

Köhler et al. (1992) points out the HSP70 in invertebrates of soil as a possible tool in the monitoring of environmental toxicants. In a study conducted by Zanger et al. (1996), the authors exposed adult of the diplopod *Julus scandinavius* to substrates contaminated with different concentrations of cadmium and investigated the expression levels of HSP70. The analyses showed that an increase in the concentration of cadmium in the animals diet resulted in high levels of HSP70.

According to Zanger et al. (1996), the so-called hepatic cells of diplopods can present a strong expression of HSP70 in stress conditions induced by heavy metals. These cells are found disperse through the cells of the "fat body" layer that compose the midgut and one of their functions is to help in the detoxification of the organism (Hopkin & Read, 1992).

In a study carried out by Nadeau et al. (2001), it was investigated the feasibility of using HSP70 as marker of the presence of soil toxicants in the midgut of the earthworm *L. terrestris*. The authors concluded that HSP70 can be efficiently used in the assessment of the toxicity of soils using test organisms exposed under laboratory conditions.

5. CONCLUSION

Due to the constant release of harmful substances into the terrestrial environment, it is necessary to know their action on the organisms present there, in order to avoid triggering a possible unbalance in the ecosystems. In this sense, we tried to present potential bioindicators and biomarkers for ecotoxicological analyses of the soil. It is important to highlight that the choice of the bioindicator organism is essential to the success of environmental monitoring. Moreover, the combined use of morphological, biochemical and genotoxic methods in the evaluation of injuries in sentinel species is interesting since it provides a more complete understanding of the action of contaminants on organisms exposed, besides providing a greater reliability to the results obtained in the researches. We believe that understanding the importance of the combined use of different methodologies in assessing the toxicity of substrates will be highly beneficial for the future work of researchers in Ecotoxicology.

REFERENCES

1. T. N. Abe, R. B. Souza, C. S. Fontanetti, 2010Alterações do corpo gorduroso do diplópodo Rhinocricus padbergi após exposição a lodo de esgoto, Proceedings of XXII Congresso de Iniciação Científica da UNESP, Rio Claro, Brasil, September 2010

2. R. K. Achazi, C. A. M. Van Gestel, 2003Uptake and Accumulation of PAHs by Terrestrial Invertebrates, in: PAHs: An Ecotoxicological Perspective, Douben, P. E. T., 173 188John Wiley & Sons, 0-47156-024-3UK.

3. B. J. Alloway, 1994 Heavy metals in soil.Springer, 978-0-75140-198-1New York.

4. E. A. Almeida, 2003 Avaliação de variações bioquímicas em moluscos bivalves em resposta ao estresse ambiental. Tese de Doutorado, Universidade de São Paulo, São Paulo.

5. F. S. Amaral, A. S. Rodrigues, 2005 Metal accumulation and apoptosis in the alimentary canal of Lumbricus terrestris as a metal biomarker 18 199 206 1572-8773

6. M. M. Andréa, 2010O uso de minhocas como bioindicadores de contaminação de solos. Acta Zoológica Mexicana, (n.s.) 2, 95 107 0065-1737

7. M. Backor, B. Pawlik-Skowronska, J. Bud´ová, T. Skowronski, 2007Response to copper and cadmium stress in wild-type and copper tolerant strains of the lichen alga Trebouxia erici: metal accumulation, toxicity and non-protein thiols. Plant Growth Regulation, 52 1 17 27 0167-6903

8. L. B. S. Barreiros, J. M. David, J. P. David, 2006 Estresse oxidativo: relação entre geração de espécies reativas e defesa do organismo Química Nova, 29 1 113 123 0100-4042

9. G. Bengtsson, S. Rundgren, 1984 Ground-living invertebrates in metal-polluted forest soils Ambio 3 29 33 0044-7447

10. D. J. Betteridge, 2000What is oxidative stress? Metabolism, 49 3 8 0026-0495

11. J. G. E. A. Bierkens, 2000Applications and pitfalls of stressproteins in biomonitoring. Toxicology, 153 61 72 0030-0483X.

12. Bispo, M. J. Jourdain, M. Jauze, 1999Toxicity and genotoxicity of industrial soils polluted by polycyclic aromatic hydrocarbons (PAHs).Org. Geochem. 30 947 952 0146-6380

13. R. Bocchetti, D. Fattorini, B. Pisanelli, S. Macchia, L. Oliviero, F. Pilato, D. Pellegrini, F. Regoli, 2008Contamination accumulation and biomarker responses in caged mussels, Mytilus galloprovincialis, to evaluate bioavailability and toxicological effects of remobilized chemicals during dredging and disposal operations in harbour areas. Aquatic Toxicology, 89 257 266 0016-6445X.

14. R. H. Burdon, 1986 Heat shock and the heat shock proteins. TheBiochemical Journal, 240 2 313 324 0264-6021

15. V. Calabrese, R. Lodi, C. Tonon, V. D. Agata, M. Sapienza, G. Scapagnini, A. Mangiameli, G. Pennisi, A. M. G. Stella, D. A. Butterfield, 2005 Oxidative stress, mitochondrial dysfunction and cellular stress response in Friedreich's ataxia. Journal of the Neurological Sciences, 233 1-2 145 162 0002-2510X.

16. C. Chang, T. D. Hinesly, T. E. Bates, H. E. Doner, R. H. Dowdy, J. A. Ryan, 1987Effects of long term sludge application on accumulation of trace elements by crops, in: Land application of sludge: food chain implications, Page, A.L.; Logan, T.J. & Ryan, J.A. 53 66Lewis Publishers, 0-87371-083-5

17. A. Christofoletti, C. S. Fontanetti, 2010Evaluación de la genotoxicidad de los lodos de depuradora y lavinaza por sistema experimental Allium cepa, Proceedingsof VIII Congresso de laAssociación de Latinoamericana de Mutagénesis, Carcinogenésis y Teratogenésis Ambiental (ALAMCTA), Viñadel Mar, Chile, October 2010.

18. L. D. Claxton, G. A. Umbuzeiro, D. M. Demarini, 2010 The Salmonella mutagenicity assay: The stethoscope of genetic toxicology for the 21st century Environmental Health Perspectives, 11 11 1515 1522 0091-6765

19. J. D. Cogo, A. F. Siqueira, A. C. Ramos, Z. M. A. Cruz, A. G. Silva, 2009Utilização de enzimas do estresse oxidativo como biomarcadoras de impactos ambientais. Natureza on line, 7 1 37 42 1806-7409

20. G. P. Creissen, E. A. Edwards, P. M. Mullineaux, 1994Glutathione reductase and ascorbate peroxidase, in: Causes of photoxidative stress and amelioration of defence systems in plants. Foyer, C.H. & Mullineaux, P.M, CRC Press: Boca Raton, 0-84935-443-9

21. R. Dallinger, F. Prosi, 1988 Heavy metals in the terrestrial isopod Porcellio scaber Latreille. II. Subcellular fractionation of metal-accumulating lysosomes from hepatopancreas. Cell Biol Toxicol,4 97 109 0742-2091

22. R. Dallinger, 1993Strategies of metal detoxification in terrestrial invertebrates, in: Ecotoxicology of Metals in Invertebrates. Dallinger, R. & Rainbow, P.S. 246 289Boca Raton: Lewis Publishers, 0-87371-734-1

23. M. H. Depledge, 1993The rational basis for the use of biomarkers as ecotoxicological tools, in: Nondestructuve biomarkers in vertebrates. Fossi, M.C. & Leonzio, C., 271 285Lewis Publichers Boca Raton, 0-87371-648-5

24. M. H. Depledge, J. J. Amaral-Mendes, B. Daniel, R. S. Halbrook, P. Kloeppersams, M. N. Moore, D. B. Peakall, 1993The conceptual basis of the biomarker approach, In: Biomarker: Research and application in the assessment of environmental health, Peakall, D.B. & Shugart, L.R., 15 29NATO ASI Series, Berlin

25. X. Diao, J. Jensen, A. D. Hansen, 2007 Toxicity of the anthelmintic abamectin to four species of soil invertebrates Environmental Pollution 148 514 519 0269-7491

26. T. A. Diplock, J. L. Charleux, G. Crozier-Willi, F. J. Kok, C. Rice-Evans, M. Roberfroid, W. Stahl, J. Viña-Ribes, 1998 Functional food science and defense against reactive oxidative species British journal of nutrition, 80 77 112 0007-1145

27. R. Eisler, 1986 Dioxin hazards to fish, wildlife, and invertebrates: a synoptic review.U.S. Fish and Wildlife Service Biological Report. 85(1.8).

28. Environmental Protection Agency- EPA. 1986 Quality criteria for water.EPA 440/5- 86-001.US Environmental Protection Agency, Washington, D.C.

29. B. C. Faulkner, R. L. Lochmiller, 2000 Increased abundance of terrestrial isopod populations in terrestrial ecosystems contaminated with petrochemical wastes. Arch. Environ. Contam. Toxicol. 39 1 86 90 0090-4341

30. T. C. C. Fernandes, D. E. C. Mazzeo, M. A. Marin-Morales, 2007 Mechanism of micronuclei formation in polyploidizated cells of Allium cepa exposed to

trifluralin herbicide Pesticide Biochemistry and Physiology 88 252 259 0048-3575

31. T. C. C. Fernandes, D. E. C. Mazzeo, M. A. Marin-Morales, 2009 Origin of nuclear and chromosomal alterations derived from the action of an aneugenic agent Trifluralin herbicide Ecotoxicology and Environmental Safety 72 1680 1686 0147-6513

32. C. S. Fontanetti, M. I. Camargo-Mathias, B. Tiritan, 2004The fat body in Rhinocricus padbergi (Diplopoda, Spirobolida). Lheringia. 94 4 351 355 0073-4721

33. C. S. Fontanetti, C. A. Christofoletti, T. G. Pinheiro, T. S. Souza, J. Pedro-Escher, 2010Microscopy as a tool in toxicological evaluations, In: Microscopy: Science, Technology, Applications and Education. Méndez-Vilas, A. & Diaz, J, 1001 1007Badajoz: Formatex Research Center, 978-8-46146-190-5

34. C. S. Fontanetti, B. Tiritan, M. I. Camargo-Mathias, 2006Mineralized badies in the fat body of Rhinocricus padbergi (DIPLOPODA). Braz. J. morphol. Sci. 23 3-4 487 493 0102-9010

35. D. R. J. Freitas, Junior. I. S. Vaz, A. Masuda, 2008Expressão e atividade enzimática de glutationa-S-transferase em tecidos de fêmeas de Boophilus microplus. Revista Brasileira de Parasitologia Veterinária, 17 2 99 104 0010-3846X.

36. J. A. P. Godoy, C. S. Fontanetti, 2010 Diplopods as Bioindicators of Soils: Analysis of Midgut of Individuals Maintained in Substract Containing Sewage Sludge WaterAir & Soil Pollution, 210 389 398 0167-6369

37. S. Gräff, M. Berkus, G. Alberti, H. R. Köhler, 1997 Metal accumulation strategies in saprophagous and phytophagous soil invertebrates: a quantitative comparison BioMetals 10 45 53 0966-0844

38. W. F. Grant, 1994The present status of higher plant bioassays for the detection of environmental mutagens. Mutation Research, 310 2 175 185 1383-5718

39. P. Greenslade, 2007The potencial of Collembolla to act as indicators of landscape stress in Australia. Australian Journal of Experimental Agriculture. 47 424 434 1836-0939

40. C. K. Grisolia, 2005 Agrotóxicos: Mutações, Câncer e Reprodução. Editora Universidade de Brasília, ISBN 852300808X, Brasília.

41. B. Hamer, D. P. Hamer, W. E. G. Müller, R. Batel, 2004Stress-70 proteins in marine mussel Mytilus galloprovincialis as biomarker of environmental pollution: a field study. Environment International, 30 873 882 0160-4120

42. Heikens, W. J. G. M. Peijnenburg, A. J. Hendriks, 2201Bioaccumulation of heavy metals in terrestrial invertebrates.Environmental Pollution. 113 3 385 393 0269-7491

43. Helling, S. A. Reinecke, A. J. Reinecke, 2000Effects of the Fungicide Copper Oxychoride on the Growth and Reproduction of Eisenia fetida (Oligochaeta). Ecotoxicology and Environmental Safety, 46 108 116 0147-6513

44. D. Hodkinson, J. K. Jackson, 2005 Terrestrial and aquatic invertebrates as bioindicators for environmental monitoring, with particular reference to mountain ecosystems. Environmental Management 35 5 649 666 0036-4152X.

45. S. P. Hopkin, H. J. Read, 1992 The biology of millipedesOxford University Press, 233 0-19857-699-4York.

46. S. P. Hopkin, 1989 Ecophysiology of metals in terrestrial invertebrates Elsevier Applied Science, 1-85166-312-6

47. S. P. Hopkin, 1990 Critical concentrations, pathways of detoxification and cellular ecotoxicology ofmetals in terrestrial arthropods. Functional Ecology 4 3 321 327 0269-8463

48. S. P. Hopkin, K. Watson, Martin, M. H. , M. L. Mould, 1985The assimilation of heavy metals by Lithobius variegatus and Glomeris marginata (Chilopoda; Diplopoda). Bijdragen tot de dierkunde. 55 1 88 94 0067-8546

49. M. Hubert, 1979Localization and identification of mineral elements and nitrogeneous waste in Diplopoda, In: Myriapod Biology, CAMATINI, M., 127 134Academic Press, 0-12155-750-2

50. R. J. S. Jacques, F. M. Bento, Z. I. Antoniolli, F. A. O. Camargo, 2007Biorremediação de solos contaminados com hidrocarbonetos aromáticos policíclicos. Ciência Rural. 37 4 1192 1201 0103-8478

51. C. C. Júnior, M. O. Marques, L. C. T. Júnior, 2008Efeito residual de quatro aplicações anuais de lodo de esgoto e vinhaça na qualidade tecnológica da cana-de-açúcar. Eng. Agric. 28 1 196 203 0100-6916

52. R. D. Kale, 1998Annelids. in: Applied Soil Biology and Ecology. Veeresh, G.K. & Rajagopal, D. 90 109Oxfordand IBH Publishing Co. Pvt. Ltd., New Dehli, 0-83642-291-0

53. J. E. Kammenga, R.. Dallinger, M. H. Donker, H. R. Köhler, V. Simonsen, R. Triebskorn, J. M. Weeks, 2000 Biomarkers in terrestrial in terrestrial invertebrates for ecotoxicological soil risks assessment. Reviews of Environmental Contamination and Toxicology, 164 93 147 0179-5953

54. S. Knasmüller, E. Gottmann, H. Steinkellner, A. Fomin, C. Pickl, A. Paschke, R. Göd, M. Kundi, 1998 Detection of genotoxic effects of heavy metal contaminated soils with plant bioassays. Mutation Research 420 37 48 1383-5718

55. H. R. Köhler, G. Alberti, 1992The effect of heavy metal stress on the intestine of diplopods, in: Advances in Myriapodology. Meyer, E.; Thaler, K. & Schedl, W. 257 267 10Innsbruck: Ber nat-med, ISBN 370300245X.

56. H. R. Köhler, (2002, 2002 Localization of metals in cells of saprophagous soil arthropods (Isopoda Diplopoda, Collembola).Microscopy Research and Technique 56 393 401 1097-0029

57. H. R. Köhler, R. Triebskorn, 1998 Assessment of the cytotoxic impact of heavy metals on soil invertebrates using a protocol integrating qualitative and quantitative components Biomarkers 3 2 109 127 0135-4750X.

58. h. R. Köhler, k. Hüttenrauch, m. Berkus, s. Gräff, g. Alberti, 1996aCellular hepatopancreatic reactions in Porcellio scaber (Isopoda) as biomarkers for the evaluation of heavy metal toxicity in soils. Applied Soil Ecology, 3 1 15 0929-1393

59. H. R. Köhler, B. Rahman, S. Gräff, M. Berkus, R. Triebskorn, 1996bExpression of the stress-70 protein family (hsp70) due to heavy metal contamination in the slug, Deroceras reticulatum: an approach to monitor sublethal stress conditions. Chemosphere, 33 7 1327 1340 0045-6535

60. H. R. Köhler, R. Triebskorn, W. Stöcker, P. M. Kloetzel, G. Alberti, 1992 The 70 kD heat shock protein (hsp 70) in soil invertebrates: a possible tool for monitoring environmental toxicants. Archives of environmental contamination and toxicology 22 334 338 0090-4341

61. O. Kovalchuk, et al. 1998 The Allium cepa chromosome Aberration Test Reliably Measures Genotoxicity of Soils of Inhabited Areas in the Ukraine

Contaminated by the Chernobyl Accident. Mutation Research 415 12 47 57 1383-5718

62. B. Lah, T. Vidic, E. Glasencnik, T. Cepelijnik, G. Gorjanc, R. Marinsek-Logar, 2008 Genotoxicity evaluation of water soil leachates by Ames test, comet assay, and preliminary Tradescantia micronucleus assay. Environmental Monitoring and Assessment 139 107 118 0167-6369

63. D. M. Leme, M. A. Marin-Morales, 2009 Allium cepa test in environmental monitoring: A review on its application. Mutation Research 682 71 81 1383-5742

64. J. C. Lopes, L. G. Ribeiro, M. G. Araújo, M. R. B. S. Beraldo, 2005Produção de alface com doses de lodo de esgoto. Horticultura Brasileira 23 1 143 147 0102-0536

65. M. R. C. C. Lyra, M.M. ROLIM, J. A. A. SILVA, 2003Toposseqüência de solos fertirrigados com vinhaça: contribuição para a qualidade das águas do lençol freático. Revista Brasileira de Engenharia Agrícola e Ambiental. 7 3 525 532 1807-1929

66. T. H. et, al, 1995The Improved Allium/Vicia Root Tip Micronucleus Assay for Clastogenicity of Environmental Pollutants. Mutation Research, 334 2 185 195 1383-5718

67. T. H. , C. Xu, S. Liao, H. Mcconnell, B. S. Jeong, C. D. Won, 1996In situ monitoring with the Tradescantia bioassays on the genotoxicity of gaseous emissions form a closed landfill site and an incinerator. Mutation Research, 359 39 52 DOIS0165-1161(96)90008-4.

68. D. P. Magalhães, Filho. A. S. Ferrão, 2008A ecotoxicologia como ferramenta no biomonitoramento de ecossitemas aquáticos. Oecologia Brasiliensis, 12 3 355 381 1981-9366

69. B. Majer, T. Grummt, M. Uhl, S. Knasmüller, 2005 Use of plant bioassays for the detection of genotoxins in the aquatic environment Acta Hydrochimica et Hydrobiologica 33 1 45 55 0323-4320

70. O. Malaspina, E. C. M. Silva-Zacarin, 2006Cell markers for ecotoxicological studies in target organs of bees. Braz. J. morphol. Sci., 23 3-4 129 136 0102-9010

71. D. Malmezat, P. Breuille, P. Capitan, P. Miranda, C. Obled, 2000 Glutathione turnover is increased during the acute phase of sepsis in rats. Journal of Nutrition, 130 5 1239 1246 0022-3166

72. P. Mantecca, G. Vailati, R. Bacchetta, 2006 Histological changes and micronucleus inductions in the Zebra mussel Dreissena polymorpha after Paraquat exposure. Histology and histopathology 21 829 840 0213-3911

73. D. E. C. Mazzeo, C. E. Levy, D. F. Angelis, M. A. Marin-Morales, 2010 BTEX biodegradation by bactéria from effluents of petroleum refinery Science of the Total Environment 408 20 4334 4340 0048-9697

74. S. P. Mc Grath, 1887Long-term studies of metal transfers following applications of sewage sludge, In Pollutant Transport and Fate in Ecosystems, COUGHTREY, P.J.; MARTIN, M.H. & UNSWORTH, M.H., 301 317Blackwell, 0-63201-627-2UK

75. T. N. Meyer, A. L. Silva, 1999 Resposta celular ao estresse Revista da Associação Médica Brasileira, 45 2 181 188 0104-4230

76. T. R. Meyers, J. D. Hendricks, 1985Histopathology. in: Fundamental of Aquatic Toxicology: Methods and Applications. Rand, G.M., Petrocelli, S.R., 283 331Hemisphere Pub, 978-0-89116-302-2Washington.

77. S. Mishra, S. Srivastava, R. D. Tripathi, R. Kumar, C. S. Seth, D. K. Gupta, 2006 Lead detoxification by coontail (Ceratophyllum demersum L.) involves induction of phytochelatins and antioxidant system in response to its accumulation Chemosphere 65 6 1027 1039 0045-6535

78. M. Misik, T. -H, A. Nersesyan, S. Monarca, J. K. Kim, S. Knasmueller, 2011 Micronucleus assays with Tradescantia pollen tetrads: an update Mutagenesis 26 1 215 221 1464-3804

79. M. Monari, J. Foschi, R. Rosmini, M. G. Marin, G. P. Serrazanetti, 2011 Heat shock protein 70 response to physical and chemical stress in Chamelea gallina Journal of Experimental Marine Biology and Ecology 397 71 78 0022-0981

80. D. Nadeau, S. Corneau, I. Plante, G. Morrow, R. M. Tanguay, 2001 Evaluation for Hsp70 as a biomarker of effect of pollutants on the earthworm Lumbricus terrestrisCell Stress & Chaperones, 6 2 153 163 1355-8145

81. A. Nascimento, S. A. Pereira, M. B. N. L. Leite, 2008Biomarcadores como instrumentos preventivos de poluição, in: Ecotoxicologia: Princípios e Aplicações. Zagatto, P.A. & Betoletti, E, 413 432RiMa, 978-8-57656-090-6São Carlos.

82. M. Nasiruddin, A. J. Mordue, 1993 The effect of azadirachtin on the midgut histology of the locusts, Schistocerca gregaria and Locusta migratoria. Tissue and Cell. 25 6 875 884 0040-8166

83. D. P. Netto, J. C. Moreira, A. E. X. O. Dias, G. Arbilla, L. F. V. Ferreira, A. S. Oliveira, J. Barek, 2000Avaliação da contaminação humana por hidrocarbonetos policíclicos aromáticos (HPAs) e seus derivados nitrados (NHPAs): Uma revisão metodológica. Química Nova. 23 6 765 773 0100-4042

84. J. C. Niemeyer, Santos. V. C. Dos, J. M. L. Rodrigues, Silva. Da, E. M. , 2006 Comportamento de Cubaris murina Brandt (Crustacea: Isopoda) em Solo com Glifosato: Teste de Fuga em Laboratório. Journal of the Brazilian Society Ecotoxicology, 1 1 13 16 1809-8401

85. J. C. Niemeyer, V. C. Santos, P. B. Araújo, E. M. Silva, 2009Reproduction of Cubarismurina (Crustacea: Isopoda) under laboratory conditions and its use in ecotoxicity tests. Braz. J. Biol. 69 1 137 142 1519-6984

86. J. C. Niemeyer, D. Vilaça, Silva. E. M. Da, 2006Efeitos na Biomassa de Cubaris murina Brandt (Crustacea: Isopoda) Expostos ao Solo com Glifosato em Laboratório. Journal of the Brazilian Society Ecotoxicology, 1 1 17 20 0000-1809

87. L. R. Nogarol, C. S. Fontanetti, 2010 Acute and subchronic exposure of diplopods to substrate containing sewage mud: Tissular responses of the midgut Micron 41 239 246 0968-4328

88. L. R. Nogarol, C. S. Fontanetti, 2011 Ultrastructural Alterations in the Midgut of Diplopods after Subchronic Exposure to Substrate Containing Sewage Mud WaterAir & Soil Pollution, 128 539 547 0049-6979

89. J. Nordberg, E. S. J. Arnér, 2001 Reactive oxygen species, antioxidants, and the mammalian thioredoxin system. Free Radical Biology & Medicine 31 11 1287 1312 0891-5849

90. J. P. Odendaal, A. J. Reinecke, 2003 Quantifying histopathological alterations in the hepatopancreas of the woodlouse Porcellio laevis (Isopoda) as a biomarker of cadmium exposure Ecotoxicology and Environmental Safety, 56 2 319 325 0147-6513

91. D. S. Page, P. D. Boehm, G. S. Douglas, A. E. Bence, W. A. Burns, P. G. Mankiewicz, 1999Pyrogenic polycyclic aromatic hydrocarbons in sediments

record past human activity: A case study in Prince Willian Sound, Alaska. Marine Pollution Bulletin. 38 247 266 0002-5326X.

92. M. Pawert, R. Triebskorn, S. Gräff, 1996 Cellular alterations in collembolan midgut cells as a marker of metal exposure: ultrastructure and intracellular metal distribution. Science of the Total Environment, 181 187 200 0048-9697

93. M. B. Pedersen, C. A. M. van Gestel, N. Elmegaard, 2000Effects of copper on reproduction of two Collembolan species exposed through soil, food and water. Environ Toxicol Chem. 19 2579 2588 1552-8618

94. J. Pedro-Escher, C. S. Fontanetti, 2010O potencial citotóxico, genotóxico e mutagênico da vinhaça no organismo teste Allium cepa, Proceedingsof VIII Congresso de la Associación de Latinoamericana de Mutagénesis, Carcinogenésis y Teratogenésis Ambiental (ALAMCTA), Viñadel Mar, Chile, October 2010.

95. D. G. Perez, C. S. Fontanetti, 2011a Assessment of the Toxic Potential of Sewage Sludge in the Midgut of the Diplopod Rhinocricus padbergi WaterAir & Soil Pollution, 128 437 444 0049-6979

96. D. G. Perez, C. S. Fontanetti, 2011b Hemocitical responses to environmental stress in invertebrates: a review Environmental Monitoring and Assessment 177 437 447 0167-6369

97. J. Rank, M. H. Nielsen, 1998 Genotoxicity testing of wastewater sludge using the Allium cepa anaphase-telophase chromosome aberration assay. Mutation Research Genetic Toxicology and Environmental Mutagenesis. 418 2-3October, 1998), 113 119 1383-5718

98. F. Regoli, M. Nigro, E. Orlando, 1998 Lysosomal and antioxidant responses to metals in the Antarctic scallop Adamussium colbeckiAquatic Toxicology, 40 375 392 0016-6445X.

99. J. Reinecke, R. G. Nash, 1984Toxicity of 2,3,7,8-TCDD and short-term bioaccumulation by earthworms (Oligochaeta). Soil Biol. Biochem. 16 45 49 0038-0717

100. E. Riser-Roberts, 1998 Remediation of petroleum contaminated soil: biological, physical, and chemical processes.CRC Press, 978-0-87371-858-5Boca Raton, Fl.

101. M. C. Rocha, R. Shirota, 1999Disposição final de lodo de esgoto. Revista de estudos ambientais.1 3 1 25 1516-3911

102. G. S. Rodrigues, T. H. , D. Pimentel, L. H. Weinstein, 1997 Tradescantia Bioassays as monitoring systems for environmental mutagenesis: a review Critical Reviews Science, 16 325 359 0735-2689

103. J. Römbke, M. Garcia, 2000 Assessment of ecotoxicological effects of pesticides on the soil fauna and soil process under tropical conditions, Proceeding of German-Brazilian Workshop on Neotropical Ecosystems-Achievements and Prospects of Comparative Research, ISBN 300010691X, Hamburg, September 2000.

104. Júnior. L. Rover, N. F. Höehr, A. P. Vellasco, L. T. Kubota, 2001 Sistema antioxidante envolvendo o ciclo metabólico da glutationa associado a métodos eletroanalíticos na avaliação do estresse oxidativo. Química Nova, 24 1 112 119 0100-4042

105. R. Sahai, N. Shukla, S. Jabeen, P. K. Saxena, 1985Pollution effect of distillery on the growth behavior of Phaseolus radiates L. Environmental Pollution. 37 245 243 0269-7491

106. C. Schlatter, 1994Chlorinated dibenzo-p-dioxins and-furans: problems in analysis of biomarkers. Clin. Chem., 40 7 1405 1408 0009-9147

107. G. A. Schultz, 1982Isopoda, in: Synopsis and classification of living organisms. Parker, S.P. 249 254McGraw-Hill, 0-07079-031-0York.

108. N. Shima, L. Z. Xiao, F. Sakuramoto, S. Ichikawa, 1997 Young inflorescence-bearing shoots with roots of Tradescantia clone BNL 4430 cultivated on nutrient solution circulating systems: an alternative to potted plants and cuttings for mutagenicity tests. Mutation Research 395 199 208 1383-5718

109. J. Silva, V. Heuser, V. Andrade, 2003Biomonitoramento Ambiental, In: Genética Toxicológica, Silva, J.; Erdtmann, B. & Henriques, J.A.P. 167 178Alcance, 8-57592-011-1Alegre.

110. E. C. M. Silva-Zacarin, A. Gregorc, Moraes. R. L. M. Silva de, 2006 In situ localization of heat-shock proteins and cell death labelling in the salivary gland of acaricide-treated honeybee larvae Apidologie 37 507 516 0044-8435

111. S. Sinkkonen, T. Rantio, A. Vattulainen, J. P. Aittola, J. Paasivirta, M. Lahtipera, 1995 Chlorohydrocarbons, PCB congerners, polychorodioxins, furans and dibenzothiophenes in pine needles in the vicinity of a metal reclamation plant Chemosphere 30 2227 2239 0045-6535

112. J. P. Sousa, M. M. Gama, C. Pinto, A. Keating, F. Calhôa, M. Lemos, C. Castro, T. Luz, P. Leitão, S. Dias, 2004 Effects of land-use on Collembola diversity patterns in a Mediterranean landscape Pedobiologia 48 609 622 0000-0031

113. T. S. Souza, C. S. Fontanetti, 2011 Morphological biomarkers in the Rhinocricus padbergi midgut exposed to contaminated soil Ecotoxicology and Environmental Safety 74 10 18 1090-2414

114. T. S. Souza, D. F. Angelis, C. S. Fontanetti, 2011 Histological and Histochemical Analysis of the Fat Body of Rhinocricus padbergi (Diplopoda) Exposed to Contaminated Industrial Soil WaterAir & Soil Pollution. DOIs11270-011-0786-9.

115. T. S. Souza, F. A. Hencklein, D. F. Angelis, R. A. Gonçalves, C. S. Fontanetti, 2009The Allium cepabioassay to evaluate landfarming soil, before and after the addition of rice hulls to accelerate organic pollutants biodegradation. Ecotoxicology and Environmental Safety.72 1363 1368 0147-6513

116. C. A. Spadotto, M. A. F. Gomes, L. C. Luchini, M. M. Andréa, 2004 Monitoramento do risco ambiental de agrotóxicos: princípios e recomendações. Embrapa Meio Ambiente, 42Jaguariúna, 1516-4691

117. K. Stanek, D. Drobne, P. Trebs, 2006 Linkage of biomarkers along levels of biological complexity in juvenile and adult diazinon fed terrestrial isopod (Porcellioscaber, Isopoda, Crustacea). Chemosphere 64 1745 1752 0045-6535

118. H. Steinkellner, F. Kassie, S. Knasmüller, 1999 Tradescantia-micronucleus assay for the assessment of the clastogenicity of Austrian water. Mutation Research 426 113 116 0027-5107

119. R. D. Stephens, M. X. Petreas, D. G. Hayward, 1995 Biotransfer and bioaccumulation of dioxins and furans from soil: chickens as a model for foraging animals The Science of the Total Environment 175 253 273 0048-9697

120. L. E. Sverdrup, A. E. Kelley, P. H. Krogh, T. Nielsen, J. Jensen, J. J. Scott-Fordsmand, J. Stenersen, 2001Effects of eight polycyclic aromatic compounds on the survival and reproduction of the springtail FolsomiafimetariaL. (Collembola, isotomidae).Environ. Toxicol. Chem. 20 6 1332 1338ISSN 07307268.

121. T. M. Tavares, F. M. Carvalho, 1992Avaliação de exposição de populações humanas a metais pesados no ambiente: exemplos do Recôncavo Baiano. Química Nova. 15 2 147 154 0100-4042

122. R. Triebskorn, I. F. Henderson, A. P. Martin, 1999Detection of iron in tissues from slugs (Deroceras reticulatum Müller) after ingestion of iron chelates by means of energy-filtering transmission electron microscopy (EFTEM). Pesticide Science, 55 55 61 0003-1613X.

123. R. Triebskorn, H. R. Köhler, T. Zanh, G. Vogt, M. Ludwing, S. Rumpf, M. Kratzmann, G. Alberti, V. Storch, 1991Invertebrate cells as targets for Hazardous substances. Zeitschrift fuer angewandte Zoologie. 78 277 287ISSN00442291.

124. G. A. Umbuzeiro, V. M. F. Vargas, 2003Teste de mutagenicidade com Salmonella typhimurium (Teste de Ames) como indicador de carcinogenicidade em potencial para mamíferos, in: Mutagênese Ambiental. Ribeiro, L. R.; Salvador, D. M. F. & Marques, E. K. 81 112ed. ULBRA, 8-57528-067-8

125. United State Environmental Protection Agency- USEPA. 2003Draft Dioxin Reassessment, National Academy of Sciences (NAS) Review Draft 2003. U.S. Environmental Protection Agency, Washington, DC.

126. United State Environmental Protection Agency- USEPA. 2000Method 4425: screening extracts of environmental samples for planar organic compounds (PAHS, PCBS, PCDDS/PCDFS) by a reporter gene on a human cell line. EPA Office of Solid Waste, SW 846 Methods, Update IVB.

127. M. Valko, C. J. Rhodes, J. Moncol, M. Izakovic, M. Mazur, 2006 Free radicals, metals and antioxidants in oxidative stress-induced cancer. Chemico-Biological Interactions, 160 1 1 40), 0009-2797

128. K. Van de Braak, 2002 Haemocytic defence in black tiger shrimp (Penaeus monodon)Doctoral Thesis, Wageningen University, 9-05808-651-8

129. K. Veltman, M. A. J. Huijbregts, M. G. Vijver, W. J. G. M. Peijnenburg, P. H. F. Hobbelen, J. E. Koolhaas, C. A. M. Van Gestel, P. C. J. Van Vliet, A. J. Hendriks, 2007 Metal accumulation in the earthworm Lumbricus rubellus. Model predictions compared to field data Environmental Pollution 146 428 436 0269-7491

130. B. C. Ventura, 2009 Investigação da mutagenicidade do azocorante comercial BDCP (Black Dye Commercial Product), antes e após tratamento microbiano, utilizando o sistema-teste de Allium cepa. Tese de Doutorado. Instituto de Biociências. Universidade Estadual Paulista. Rio Claro-SP, 205p.

131. J. Walker, 1976 Host defense mechanisms in the gastrointestinal tract. Pediatrics 57 6 901 916 1098-4275

132. P. A. White, L. D. Claxton, 2004 Mutagens in contaminated soil: a review. Mutation Research 567 227 345 1383-5742

133. J. Wölz, et al. 2011 Investigation on soil contamination at recently inundated and non-inundated sites Journal of Soils Sediments, 11 82 92 1614-7480

134. World Health Organization- WHO. 2010 Dioxins and their effects on human health.Fact sheet 225www.who.int/inf-fs/en/fact225.html. Accessed in May 2011.

135. M. Zanger, G. Alberti, M. Kuhn, H. R. Köhler, 1996 The stress-70 protein family in diplopods: induction and characterization Journal of Comparative PhysiologyB: Biochemical, Systemic, and Environmental Physiology, 165 622 627 0174-1578

Chapter 7

The Effect of Calcium Peroxide on the Phenol Oxidase and Acid Phosphatase Activity and Removal of Fluoranthene from Soil

Anna Małachowska-Jutsz, Magdalena Niesler

Environmental Biotechnology Department, Silesian University of Technology

ABSTRACT

A study has been conducted to enhance fluoranthene degradation by combining biodegradation with hydrogen peroxide oxidation, as a chemical oxidant calcium peroxide has been used. The impacts of addition of calcium peroxide on microbial activity (phenol oxidase and acid phosphatase) as well as fluoranthene removal efficiency were investigated. It was observed that in the presence of calcium peroxide, the removal efficiency of fluoranthene on day 30 of the experiment was threefold higher as compared to a reference sample. It was found that the activity of phenol oxidase was stimulated on days 1, 7, and 14, by the presence of fluoranthene, whereas stimulation of the acid phosphatase activity in the samples of soil contaminated by fluoranthene was observed only after 14 days of the experiment. This may indicate that the induction period for this enzyme is longer compared with the induction period for phenol oxidase. The inhibition of the activity of both enzymes was observed in the presence of calcium peroxide.

KEYWORDS

Fluoranthene, AOPs, Phenol oxidase, Acid phosphatase, Soil decontamination

1 INTRODUCTION

Polycyclic aromatic hydrocarbons (PAHs) are commonly occurring contaminants. Their poor solubility in water causes them to be highly absorbed by soil and bottom sediments. This makes their accessibility limited for microorganisms responsible for the biodegradation process. Fluoranthene which is one of 16 priority pollutant PAHs according to Environmental Protection Agency consists of naphthalene and benzene unit connected by a five-membered ring. PAHs having four or more aromatic rings, e.g., fluoranthene, are more recalcitrant to microbial degradation. Fluoranthene is the most common and persistant PAHs pollutants and is therefore frequently chosen as a model compound for ex situ biodegradation studies (Sepic et al.2003). When the efficacy of biodegradation is insufficient and the rate of natural degradation of the contaminants is too low, then it is necessary to use additional solutions. Chemical oxidation is a promising degradation method for an extensive variety of persistent compounds, e.g., PAHs. The recent studies demonstrated that chemical oxidation can coexist with aerobic metabolism in soils and that combination is more effective solution compared to natural bioremediation processes. In Advanced Oxidation Processes (AOPs), highly reactive free radicals are formed, especially hydroxyl radicals that have a high redox potential (2.8 V) and the capability to oxidize practically any organic compound (Biń and Zieliński 2000; Liu et al.2014). Currently, research is being done to find the possibility of using other reagents in the processes of advanced oxidation. Compounds like peroxides are of high interest. Peroxides are a group of compounds that contain a peroxide group in its structure. During the degradation of these compounds, H_2O_2 (1, 2, 3) is produced, which is the source of oxygen for microorganisms taking active part in the biodegradation of contaminants while at the same time being the source of free radicals capable of oxidizing organic contaminants.

Calcium peroxide naturally decomposes very slowly to form calcium hydroxide and oxygen.

Depending upon the environment, the decomposition proceeds according to the reactions below:

$$2CaO_2 + 2H_2O \rightarrow 2Ca(OH)_2 + O_2 \quad (1)$$

or

$$CaO_2 + 2H_2O \rightarrow Ca(OH)_2 + H_2O_2 \quad (2)$$

$$2H_2O_2 \rightarrow 2H_2O + O_2 \quad (3)$$

The application of calcium peroxide has been found to be one of the most efficient for removal petroleum hydrocarbons, polycyclic aromatic hydrocarbons, tetrachloroethylene, and 2,4,6-trinitrotoluene from soil (Cassidy et al. 2008; Goi et al. 2011; Khodaweisi et al. 2011; Bianchi-Mosquera et al. 1994; Arienzo 2000; Hanh et al. 2005). Low solubility of calcium peroxide allows continuous release of hydrogen and oxygen peroxide for a long period of time (Goi et al. 2011). Moreover, the addition of calcium peroxide allows decomposition of contaminations in a wide pH range (Arienzo 2000). In order to state the full success of bioremediation, one cannot take into account only the degradation (full or partial) of the contaminant. It often occurs that during the degradation of PAHs, substances that are more toxic than the initial ones appear in the ecosystem. The progress of the bioremediation process of soils contaminated with hydrocarbons is usually monitored with the use of conventional methods of chemical analysis, which however do not evaluate the actual influence of these contaminants on the biotic elements of the environment. Based on literature review, it is known that in a simplified but reliable way, the process of bioremediation of soils contaminated with hydrocarbons can be tracked using, among others, activity of polyphenol oxidase, acid phosphatase, and dehydrogenase (Chaudhary et al. 2012). Research of the activity of the enzymes of soil microorganisms is widely used to determine biodiversity and the quality of soil (Chaudhary et al. 2012; Quasenian et al. 2012; Małachowska-Jutsz and Miksch 2010). The microbiological activity of soil depends, inter alia, on the type of contamination and the concentration of the analyzed compound (Chaudhary et al.2012). The enzymes taking active part in the degradation of polycyclic aromatic hydrocarbons, including fluoranthene, are, inter alia, oxygenase, dehydrogenase, ligninolytic enzymes (lignin peroxidase and peroxidase dependent on Mn), and phenol oxidase (Haritash and Kauskik 2009).

In these research, the process of remediation of contaminated soil was stimulated by adding calcium peroxide which been used as activator for fluoranthene degradation on laboratory scale. Two concentrations of calcium peroxide selected in previous experiment (Małachowska-Jutsz et al. 2014a, b) were examined. The influence of dose calcium peroxide on phenol oxidase and acid phosphatase as well as fluoranthene removal efficiency was investigated. Determining the activity of acid phosphatase and polyphenol oxidase was selected for the research because polyphenol oxidase may give indications on the oxidative potential of soil (Gianfreda et al. 2005), while acid phosphatase is actively involved in phosphorus metabolism in soil and is widely determined in research of the effects of various pollutants on soil microflora. The proposed approach is the combination of biological mediation methods with chemical processes, which allow to increase the removal of hydrocarbons efficiency and the reduction of processing time.

2 MATERIALS AND METHODS

The soil (loamy sand) was collected from the 0–20-cm layer of uncontaminated land Kotlarnia. The soil was dried at room temperature (22–25 °C) for 2 weeks and passed through a sieve with a mesh size of 2 mm, in order to homogenize it. An analysis of basic physicochemical parameters was performed in accordance to the methodology of Ostrowska et al. (1991) which consisted of analyzing pH, organic substance, organic carbon (Tiurin method), buffering (Arrhenius method), hydrolytic acidity (Kappen method), total nitrogen (Kjeldahl method), phosphorus (ISO 11263, 1994); the distribution of soil particle size was established by an aerometric method (PN-R-04032 1998). Several characteristics of the soil are presented in Table 1. The experiment and analysis were performed in triplicate.

Table 1. Physical-chemical parameters of soil

Parameter	Value (mean±standard deviation)
pH_{H2O}	6.33±0.17
Organic matter	12.068±0.38 g/kg
Organic carbon	7.0±0.22 g/kg
Buffer capacity	0.24±0.02 [B]
Hydrolytic acidity	0.525±0.03 [cmol (+)/kg]
Total nitrogen	0.55±0.022 g/kg
Phosphorus (P_2O_5)	0.018±0.0027 g/kg
Sand	84.05 %
Silt	2.55 %
Clay	13.4 %

Ten kilograms of soil was added into each of six containers with a volume of 20 l (each). The research was conducted in the following combinations (Table 2). Calcium peroxide (CaO_2), powder, 200 mesh (0.075 mm), purchased from Aldrich, was applied for that soil treatment. The contents of CaO_2 in calcium peroxide were 76.4 ± 0.1 % (with 17.0 ± 0.1 % of available oxygen). The soil samples were mixed with calcium peroxide. Remediation was carried out for 30 days in 22–25 °C. In this time, the soil was systematically moistened used distillate water in order to maintain moisture of 60 % water whole capacity, and it was mixed. The soil in each of six containers was daily mixed. Samples for enzyme assay and fluoranthene concentration determination were collected in four replicates, after 1, 7, 14, and 30 days of treatment.

Table 2. Experimental design

Sample	Calcium peroxide (mg/kg soil)	Fluoranthene (mg/kg soil)
1	0	0
2	0	1.5
3	120	0
4	240	0
5	120	1.5
6	240	1.5

2.1 Determination of Fluoranthene Concentration

The soil samples were dried in a dryer in a temperature of 30 °C and then passed through a sieve with a mesh size of 2 mm. Extraction of fluoranthene from the soil was done using dichloromethane (CH_2Cl_2) in an automatic Soxtherm extractor (Gerhardt brand) in a temperature of 135 °C for 3 h. Fifty grams of dried and homogenized soil was used for the extraction. The extracts were quantitatively transferred to beakers with a volume of 25 ml and were concentrated in room temperature with a gentle stream of nitrogen to a volume of about 1 ml. The concentrated extracts were cleaned on glass columns diameter of 1 cm, filled with 1 g of Florisil (deactivated with 4 % of water). The fraction containing fluoranthene was washed out of the deposit by washing it twice with dichloromethane on order to obtain 10 ml of eluate. The eluate was concentrated in room temperature using a gentle stream of nitrogen to a volume of about 1 ml. An HP1050 liquid chromatography with a HP1046 fluorescence detector from Hewlett-Packard was used to determine the content of fluoranthene. A BAKERBOND PAH 16 Plus column with Baker precolumn was used. The dimensions of the column: length 250 mm, inner diameter 3 mm. It was filled with nonpolar phase C18 with a particle size of 5 µm.

2.2 The Acid Phosphatase Activity Assay

The acid phosphatise activity was determined by the Tabatabai and Bremner method (Page 1982). One gram of soil and 0.25 ml of toluene were added into an Erlenmeyer flask and carefully mixed. The flask was tightly closed and left for 10 min. Afterward, 4 ml of MUB buffer with pH 6.5 and 1 ml of *p*-nitrophenyl phosphate sodium buffer solution were added. The flask was closed and mixed for a few seconds and then put into incubation in a temperature of 37 °C for 1 h. After the incubation, the reaction was interrupted by adding 1 ml of 0.5 M calcium chloride and 4 ml of 0.5 M sodium hydroxide. It was mixed for a few seconds, and then filtered through filter paper (Whatman no. 12) into a test tube. The absorbance of the filtrate was measured with a spectrophotometer at a wavelength of 400 nm. The concentration of *p*-nitrophenol expressed in milligrams per kilogram was calculated based on a calibration curve.

Activity of acid phosphatase expressed in mgg∗hmgg∗h was calculated using the formula:

$$\text{Activity} = \frac{released\,p-nitrophenol(mg)}{soil\,sample(g)*incubation\,time(h)}\left[\frac{mg}{g*h}\right]$$

2.3 The Phenol Oxidase Activity Assay

The phenol oxidase activity was determined by Bob Sinsabaugh method (Allison Lab Protocol 2008). Ten grams of soil and 50 ml of acetate buffer were added into 50-ml Erlenmeyer flask and stirred. The 2 ml of soil suspension was transferred into test tube. To the same test tube, 2 ml of 5 mM solution of L-DOPA was added. In parallel, a control sample was prepared which contained 2 ml of soil suspension and 2 ml of acetate buffer. The test sample was mixed thoroughly and incubated for 1 h at 20 °C and then centrifuged. The absorbance of the test sample supernatant and the control sample was measured at a wavelength of 460 nm. All the measurements were made in four replicates.

The phenol oxidase activity, expressed in μmolg∗hμmolg∗h was calculated according to the formula:

OD = absorbance of the sample − absorbance of the control sample

Activity=OD1,66∗incubation∗(drysoilsample(g)suspension(ml))[μmolg∗h]Activity=OD1,66∗incubation∗(drysoilsample(g)suspension(ml))[μmolg∗h]

2.4 Statistical Analysis and Enzyme Assay

Statistical analysis of the results included the calculation of the arithmetic mean and standard deviation. To verify the distribution normality, the Shapiro-Wilk test was applied. The significance of differences between individual samples was assessed using Student's *t* test. Differences were considered statistically significant if $p < 0.05$. Statistical analysis was performed using Statistica 10 StatSoft, Inc., software.

Results of enzyme activity were calculated as percentage of the control sample as:

A=K−BK∗100%[%]A=K−BK∗100%[%]

where:

A percentage of enzyme activity against the control sample [%]

K mean enzyme activity in control sample

B mean enzyme activity in test sample

3 RESULTS AND DISCUSSION

3.1 Fluoranthene Removal in Soil

The chemical structure of polycyclic aromatic hydrocarbons is one of the main factors, which have influence in their degradation rate. Hydrocarbons which contain two to three rings in the molecule are relatively easily biodegradable, and the decomposition rate is inversely proportional to the number of aromatic rings. In the case of hydrocarbons with four or more aromatic rings, the decomposition is very slow and often requires the presence of cometabolites. Crucial effect on the biodegradation of the compounds is the presence of microorganisms capable to degrade these substances. In the present study, it was found that control soil sample was contaminated with 247 µg/kg dry weight (DW) of fluoranthene. Many hydrocarbons, including fluoranthene, undergo degradation after a period of adaptation of microorganisms. At that time, they produce inductive enzymes. This process is influenced by, among others, the concentration of hydrocarbons as carbon sources: Too low limits or even prevents the production of these enzymes, and too high is toxic. In the research, it was found that in the samples into which 1.5 mg of fluoranthene/kg of soil was introduced, the biodegradation proceeded efficiently. After 30 days of self-cleaning process, the decreases of fluoranthene concentration in control sample and samples, to with 1.5 mg/kg fluorathene was added, were respectively 74.23 % (of 0.25 mg/kg) and 77 % (Fig. 1). In soils with calcium peroxide, higher fluoranthene concentration was determined (Fig. 1). This indicates the presence of microorganisms capable to degrade the hydrocarbon. The decomposition process is also affected by the soil moisture. Literature data indicate that PAHs are the most stable in air-dry soils (less than 1 % of water holding capacity (WHC)), and the biodegradation process runs most efficiently at a relative humidity of 60–65 % of WHC (Chiou 1989). This humidity was maintained throughout the experimental period, i.e., 30 days. In soils with low moisture sorption of PAHs is stronger than in the hydrated systems because organic compounds can be adsorbed not only by the soil organic matter but also by clay minerals (Maliszewska-Kordybach 1990). Another very important issue is the content of organic matter in the soil. Slightly loamy sand with low organic matter content (Table 1) in order to eliminate the influence of sorption process of fluorantede degradation was intentionally chosen. On the other hand, studies show that the microbial activity in soils was a critical factor governing the degradation of organic micro-pollutants (Małachowska-Jutsz et al. 2014a). The present study was conducted to analyze the effects of soil organic matter on the development of degradation potentials for polycyclic aromatic hydrocarbons (PAHs). Most of the degradation kinetics for PAHs by the indigenous microorganisms developed in soils can be fitted with the Logistic growth models. The microbial activities were relatively lower in the soils with the lowest and highest organic matter content, which were likely due to the nutrition limit and PAH sequestration. The microbial activities developed in humic acids (HA) were much higher than those developed in humic acids, which was demonstrated to be able to sequester organic pollutants stronger. The results suggested that the nutrition support and sequestration were the two major mechanisms that soil organic matter influenced the development of microbial PAHs degradation potentials (Yang et al. 2012).

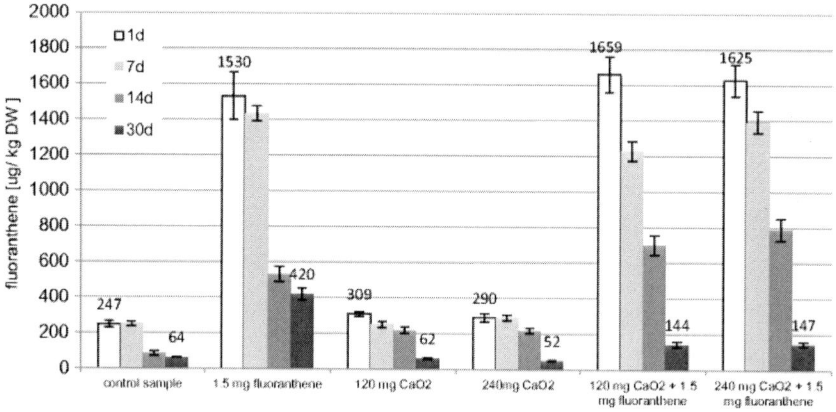

Figure. 1 Changes of fluoranthene concentration in soil samples

The addition of calcium peroxide to the soil samples, to which fluoranthene in concentration of 1.5 mg/kg of soil was added, caused an increase in fluoranthene removal. After 30 days, the removal degree of fluoranthene, in soil samples with calcium peroxide in concentration of 120 mg/kg of soil and 240 mg/kg of soil, was respectively 90.11 and 90.2 % (Fig. 1). The same phenomena was observed by Arienzo (2000). The addition of low concentrations of calcium peroxide (0.2 %) to soils contaminated with PAHs does not lead to sterilization the soil and did not cause significant changes in the metabolic activity of microorganisms. However, the additional condition was to provide the microorganisms biogenic substances in an effective amount. Generally, a molar weight ratio of carbon to nitrogen to potassium to phosphorous of from 100:10:10:10 to 100:10:1:1 is employable, with a preferred ratio of from 100:10:5:5 to 100:10:1:1, and a most preferred ratio of from 100:10:4:4 to 100:10:1 :1 (patent The bioremediation of soils containing hydrocarbon using calcium peroxide - WO 1994029242 A1, 1994). In the present study, no biogenic substances were added.

pH is an important factor influencing the degradation of PAHs. The pH was monitored in the course of the study (Table 3). The addition of calcium peroxide caused the increase of pH value from 6.33 to 6.84 for 120 mg CaO_2/kg and to 7.16 for 240 mg CaO_2/kg of soil samples. Many studies have shown that optimal conditions for microbial degradation of hydrocarbons are pH 6.5–7.5 (Małachowska-Jutsz and Miksch 2010; Gan and Ng 2012). According to Pawar (2012) bacterial populations were greater at pH 7.5 which was highly correlated with soil adenosine triphosphate (ATP) levels. It was therefore evident that the greatest rates of PAHs degradation were associated with the greatest bacterial population. Soil enzyme activities in general were also greatest at pH 7.5.

Table 3. Changes of pH_{H2O}

Sample	1 day	7 days	14 days	30 days
Control sample	6.33	6.31	6.23	6.27
1.5 mg fluoranthene	6.40	6.42	6.29	6.28
120 mg CaO_2	6.84	6.72	6.78	6.79
240 mg CaO_2	7.16	6.96	7.00	7.28
120 mg CaO_2+1.5 mg fluoranthene	7.02	6.76	6.77	7.16
1240 mg CaO_2+1.5 mg fluoranthene	7.28	6.75	7.02	7.38

3.2 The Changes of Acid Phosphatase Activity During Remediation

Soil also contains a resource of available enzymes secreted from living cells of microorganisms and released into the environment in the lytic processes, which operate independently from parent cells, and their activity is controlled by the conditions in the environment (Brzezińska 2006; Dick 1997; Gianfreda et al.2005).

Acid phosphomonoesterase, commonly referred to as acid phosphatase, is actively involved in the first stage of decomposition of organic matter in the soil environment; thereby, it is involved in the processes of mineralization of organic phosphorus. The activity of phosphatases in soil reflects the activity of enzymes associated with soil colloids and humic substances, free phosphatases in the soil solution, and phosphatases associated with live and dead cells of plants and microorganisms. The main source of phosphatases in soil is primarily soil microorganisms and plant roots and soil fauna.

The effect of soil modifications on the activity of acid phosphatase is shown in Fig. 2. The addition of fluoranthene on days 1 and 7 of the experiment resulted in slight decrease of acid phosphatase activity in comparison with a control sample. Stimulation of the enzyme activity was observed in the course of the process of the remediation of the soil contaminated with fluoranthene after 14 and 30 days. This may be indicative of the contribution of this enzyme to the biodegradation process of fluoranthene and also of relatively long period of induction of this enzyme by soil microorganisms (Borràs et al. 2010). On day 30 of the study, acid phosphatase activity in the samples contaminated with fluoranthene was on average about 30 % higher than in a reference sample. The highest percentage of activity decrease of acid phosphatase (37.08 %) was observed after day 1 of the experiment in the soil sample to which CaO_2 was added at a concentration of 120 mg/kg soil after, while in the subsequent days of the experiment, the phosphatase activity increased. On day 30 of the experiment, acid phosphatase activity in the sample with the addition of calcium peroxide was almost the same as in the control. For the soil sample containing CaO_2 (240 mg/kg), as in the previous case, the highest decrease of acid phosphatase activity, compared to the control, was observed on day, while lowest on day 30 of the experiment. Addition of calcium peroxide caused an increase in pH of the soil (Table 3) and at the same time caused decrease of acid phosphatase activity in comparison with the activity of the enzyme observed in the control sample. By Bielińska (2005), optimum pH of the medium for acid phosphatase is in the range of 4–6. pH observed in this experiment was higher, what could affect on decrease of acid phosphatase activity. A similar phenomenon of acid phosphatase activity was observed by Quasenian et al. (2012), in studies of soil contaminated with anthracene, after 3 months of the experiment, they found a slight decrease in the activity of acid phosphatase.

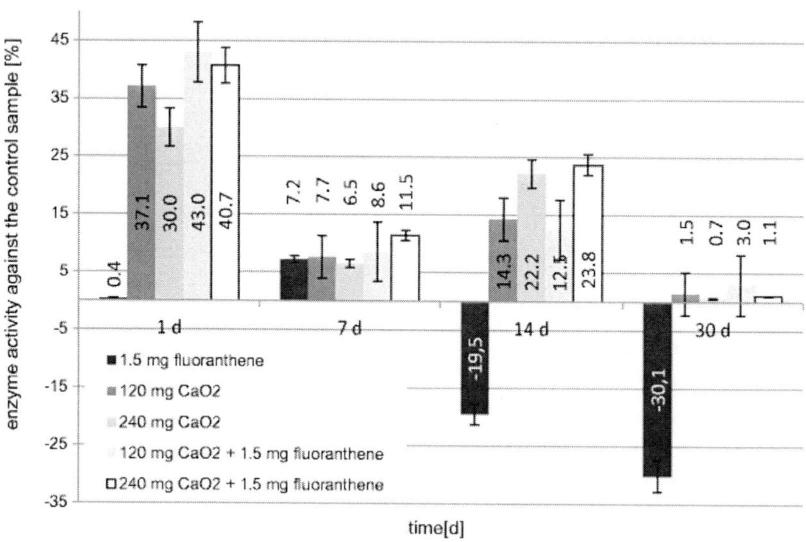

Figure. 2 The effect of fluoranthene and calcium peroxide on the activity of acid phosphatase

The addition of fluoranthene in combination with calcium peroxide concentration of 120 mg/kg DW of the soil on day 1 of the experiment caused the highest percentage of inhibition of this enzyme. The effect of the activity of acid phosphatase was sinusoidal. No significant differences between the activity of acid phosphatase and the concentration of calcium peroxide were observed.

Statistical analysis of data indicated significant effects of the applied soil modifications only on day 1 of the experiment (the exception was the sample with the addition of fluoranthene) (Table 4).

3.3 The Changes of Polyphenol Oxidase Activity During Remediation

Bacterial and fungal phenol oxidases are one of the most important enzymes involved in the degradation of polycyclic aromatic hydrocarbons (Baldrin et al. 2000; Chaudhary et al. 2012; Haritash and Kauskik 2009; Quasenian et al. 2011, 2012). These enzymes are released into the environment by secretion or cell lysis and are responsible for nonspecific catalysis of reactions such as oxidation of Mn^{2+} and Fe^{2+}, polymerization, depolymerization, or transformation of a wide range of compounds. Due to the toxicity of the phenolic derivatives, the reactions catalyzed by phenol oxidases affect the activity and composition of the soil microbial population (Sinsabaugh 2010).

The effect of the applied soil modifications on the activity of polyphenol oxidase activity is presented in Fig. 3. In the soil sample with fluoranthene, a stimulation of polyphenol oxidase was observed on days 1, 7, and 14 of the experiment, but on day 30, decrease of activity was observed compared to a control sample. The changes in the activity of this enzyme correlated with the changes of fluoranthene concentration in the soil sample. It was observed that the higher was the concentration of fluoranthene in a sample, the higher was the activity of polyphenol oxidase. This may suggest that polyphenol oxidase is actively participating in the degradation of polycyclic aromatic hydrocarbons and is immediately induced by soil microorganisms instantaneously when fluoranthene is introduced into the environment in an appropriate concentration. The same phenomena were observed by Patel et al. (2009), during the 15-day experiment of fluoranthene degradation. Acevado et al. (2011), on the contrary, found no phenol oxidase activity in a sample of soil containing polycyclic aromatic hydrocarbons.

Table 4. Changes of acid phosphatase activity during remediation of fluoranthene and statistical analysis of data

Soil samples	Acid phosphatase activity $\left[\frac{mg}{g*h}\right]$							
	1 day		7 days		14 days		30 days	
	AF±SD	s	AF±SD	s	AF±SD	s	AF±SD	s
Control sample	0.678±0.004		0.510±0.031		0.540±0.071		0.425±0.06	
1.5 mg fluoranthene	0.675±0.008	–	0.473±0.018	–	0.645±0.027	+	0.553±0.009	–
120 mg CaO_2	0.427±0.029	+	0.471±0.046	–	0.463±0.018	–	0.419±0.013	–
240 mg CaO_2	0.475±0.004	+	0.477±0.010	–	0.420±0.018	+	0.422±0.019	–
120 mg CaO_2+1.5 mg fluoranthene	0.387±0.01	+	0.466±0.010	–	0.472±0.015	–	0.413±0.011	–
120 mg CaO_2+1.5 mg fluoranthene	0.402±0.018	+	0.452±0.022	–	0.411±0.007	+	0.421±0.012	–

+ indicates that the difference in enzyme activity relative to the control was statistically significant at a 95 % confidence level. − indicates that the difference in enzyme activity relative to the control not statistically significant at a confidence level of 95 %

AF average acid phosphatase activity $\left[\frac{mg}{g*h}\right]$, *SD* standard deviation, *ts* significance of the differences between samples based on Student's *t* test

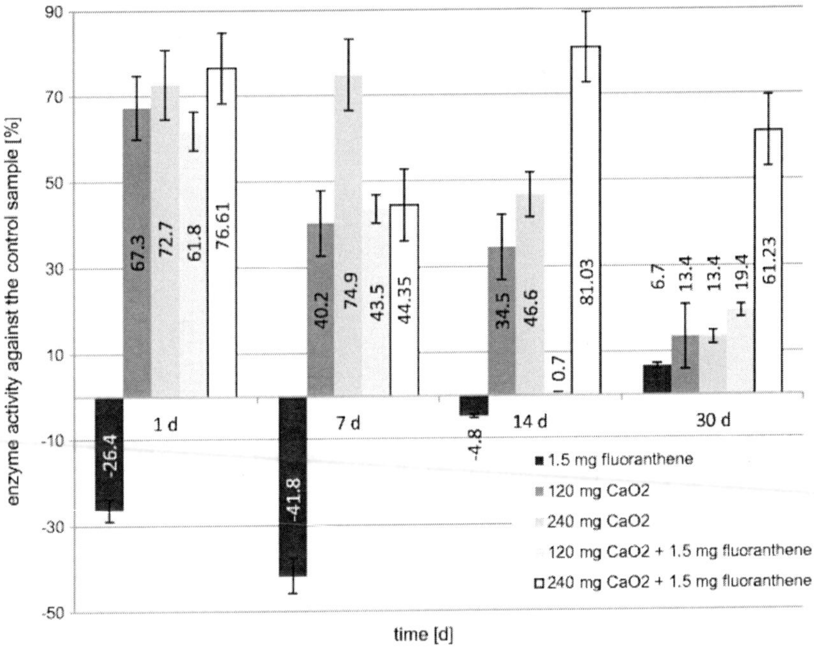

Figure. 3 The effect of fluoranthene and calcium peroxide on the activity of polyphenol oxidase

The addition of calcium peroxide to soil samples, regardless of its concentration, caused the decrease of polyphenol oxidase activity during the whole experiment. Statistical analysis showed no significant difference between the results obtained for samples with the addition of 120 mg CaO_2/kg DW of soil and 240 mg CaO_2/kg DW of soil. The highest percentage decrease of phenol oxidase activity, amounting to 67.3 and 72.68 %, was recorded on day 1 of the study. Over time, the adverse effect of the addition of calcium peroxide on the activity of polyphenol oxidase decreased. In samples with the addition of fluoranthene and calcium peroxide, the percentage of decrease of polyphenol oxidase activity was similar as in the samples containing only calcium peroxide. Statistical analysis of results showed no significant difference between the samples with the addition of calcium peroxide and those contaminated with fluoranthene and with the addition of calcium peroxide.

Statistical analysis of the obtained results in comparison with a control sample is shown in Table 5. It was shown that all of the above modifications significantly affected the result of the test in comparison with a control sample.

The exception was the sample contaminated with fluoranthene. In the course of the bioremediation process (day 14 and 30), there were no significant differences between the activity of polyphenol oxidase determined in the control sample and the sample contaminated with fluoranthene (Table5). The decrease of polyphenol oxidase activity was probably due to the formation of highly reactive free hydroxyl radicals from calcium peroxide, which can inhibit the enzyme. However, different phenol oxidases were characterized by a different optimum pH. For example, the phenol oxidases of white rot fungi have the highest activity at pH 4–6, and the phenol oxidases of brown rot fungi have the highest activity at pH 6–7.5 (Courty et al. 2009). The positive correlation (3.52) between removal of fluoranthene and polyphenol oxidase activity has been observed; however, there was no correlation (1.085) in the case of acid phosphatase (for $p = 0.05$).

4 CONCLUSIONS

The application of peroxides such as calcium peroxide allows shortening the duration of remediation of soils by increasing the content of oxygen available to the microorganisms and alkalinization of the environment. Addition of calcium peroxide to soil contaminated by fluoranthene caused approximately threefold increase in removal efficiency of this hydrocarbon on day 30 of bioremediation compared to the samples to which this reagent was not added. However, the increase of fluoranthene removal in the presence of calcium peroxide was observed, the slightly decrease of examined enzymes.

The presence of fluoranthene in soil stimulated phenol oxidase activity, which may indicate the important role of this enzyme in the decomposition of aromatic hydrocarbons.

ACKNOWLEDGMENTS

Analyses reported here were supported by grant NN523 617739 from the Ministry of Science and Higher Education. The author wishes to thank Paulina Fic for her technical support.

Table 5. Changes of polyphenol oxidase activity during remediation and statistical analysis of data

Soil samples	Acid phosphatase activity $\left[\frac{mg}{g*h}\right]$							
	1 day		7 days		14 days		30 days	
	AO±SD	ts	AO±SD	ts	AO±SD	ts	AO±SD	ts
Control sample	0.533±0.016		0.324±0.011		0.265±0.045		0.491±0.022	
1.5 mg fluoranthene	0.674±0.100	+	0.459±0.043	+	0.278±0.035	–	0.458±0.020	–
120 mg CaO_2	0.174±0.055	+	0.194±0.027	+	0.174±0.026	+	0.425±0.015	+
240 mg CaO_2	0.146±0.024	+	0.081±0.011	+	0.142±0.028	+	0.425±0.011	+
120 mg CaO_2+1.5 mg fluoranthene	0.204±0.036	+	0.183±0.014	+	0.264±0.014	–	0.395±0.016	+
120 mg CaO_2+1.5 mg fluoranthene	0.125±0.017	+	0.180±0.013	+	0.050±0.012	+	0.190±0.012	+

+ indicates that the difference in enzyme activity relative to the control was statistically significant at a 95 % confidence level. – indicates that the difference in enzyme activity relative to the control not statistically significant at a confidence level of 95 %

AO average polyphenol oxidase activity $\frac{\mu mol}{g*h}$, *SD* standard deviation, *ts* significance of the differences between samples based on Student's *t* test

REFERENCES

1. Page A. L. (1982). Methods of soil analysis Part 2. Chemical and microbiological properties. Madison, WI, USA: American Society of Agronomy, Inc. Soil Science Society of America, Inc. Publishers.

2. Acevado, F., Pizzul, L., del Pilar Castillo, M., Cuevas, R., & Diez, M. C. (2011). Degradationof polycyclic aromatic hydrocarbons by the Chilean white-rot fungus Anthracophyllum discolour. *Journal of Hazardous Materials, 185*, 212–219.

3. Allison Lab Protocol: Colorimetric Enzyme Assays, 1 (2008) by Steve Allison.

4. Arienzo, M. (2000). Use of abiotic oxidative-reductive technologies for remediation of munition contaminated soil in a bioslurry reactor. *Chemosphere, 40*(4), 441–448.

5. Baldrin, P., Wiesche, C., Gabriel, J., Nerud, F., & Zadrazil, F. (2000). Influence of cadmium and mercury on activities of ligninolytic enzymes and degradation of polycyclic aromatic hydrocarbons by *Pleurotus ostreatus*in soil. *Applied and Environmental Microbiology, 66*(6), 2471–2478.

6. Bianchi-Mosquera, G. C., Allen-King, R. M., & Mackay, D. M. (1994). Enhanced degradationof dissolved benzene and toluene using a solid oxygen-releasing compound. *Ground Water Monitoring and Remediation, 14*(1), 120–128.

7. Bielińska, E. (2005). Methods of determination of photosphatase activity (in polish). *Acta Agrophysica, Rozprawy i Monografie, 3*, 63–74.

8. Biń, A., & Zieliński, J. (2000). Chemical degradation of contaminants in industrial wastewater(in polish).*Rocznik Ochrony Środowiska, 2*, 383–404.

9. Borràs, E., Gloria Caminal, G., Sarrà, M., & Novotný, C. (2010). Effect of soil bacteria on the ability of polycyclic aromatic hydrocarbons (PAHs) removal by *Trametes versicolor* and *Irpex lacteus* from contaminated soil. *Soil Biology & Biochemistry, 42*, 2087–2093.

10. Brzezińska, M. (2006). Impact of treated wastewater on biological activity and accompanying processes in organic soils (field and model experiments) (in polish). Rozprawy i Monografie, 131. Lublin.

11. Cassidy, D.P., Northup, A., Hampton, D. (2008). The effect of AOPs on the chemical destruction of 2,4-Dinitrotoluene and on its subsequent biodegradability by Native Soil Microorganisms, 4th European Bioremediation Conference, Chania, Greece, September 3–6.

12. Chaudhary, P., Shashi, B. S., Chaudhry, S., & Nain, L. (2012). Impact of PAH on biological health parameters of soil of an Indian refinery and adjoining agricultural area – a case study. *Environmental Monitoring and Assessment, 184*, 1145–1156.

13. Chiou, C., (1989). Theoritical considerations of the portion uptake of non-ionic organic compounds by soil organic matter. In B. Sawhney, K. Brown (eds), Reactions and movement of organic chemicals in soil. Soil Science Society of America and American Society of Agronomy, Special publications, 22, (pp. 1–29).

14. Courty, P. E., Hoegger, P. J., Kilaru, S., Kohler, A., Buee, M., Garbaye, J., Martin, F., & Kues, U. (2009). Phylogenetic analysis, genomic organization, and expression analysis of multi-copper oxidasesin the ectomycorrhizal basidomycete *Laccaria bicolour. New Phytologist, 182*, 736–750.

15. Dick R. (1997). Soil enzyme activities as integrative indicators of soil health. In C. Doube, B. Gupta (Eds.), V Biological Indicators of soil Health. Pankhurst, Wallingford, USA 7 CAB International, (pp. 121–156).

16. Gan, V. S., & Ng, H. K. (2012). Current status and prospects of Fenton oxidation for the decontamination of persistent organic pollutants (POPs) in soils. *Chemical Engineering Journal, 213*, 295–317.

17. Gianfreda, L., Rao, A., Piotrowska, A., Palumbo, G., & Colombo, C. (2005). Soil enzyme activitiesas affected by anthropogenic alteractions: intensiva agfricultural practices and organic pollution. *Science of the Total Environment, 341*(1–3), 265–279.

18. Goi, A., Viisimaa, M., Trapido, M., & Munter, R. (2011). Polychlorinated biphenyls-containing electrical insulating oil contaminated soil

treatment with calcium and magnesium peroxides. *Chemosphere, 82*, 1196–1201.

19. Hanh, D. N., Rajbhandari, B. K., & Annachhatre, A. P. (2005). Bioremediation of sediments from intensive aquaculture shrimp farms by using calcium peroxide as slow oxygen release agent. *Environmental Technology, 26*(5), 581–589.

20. Haritash, A. K., & Kauskik, C. P. (2009). Biodegradation aspects of polycyclic aromatic hydrocarbons (PAHs): a review. *Journal of Hazardous Materials, 169*, 1–15.

21. ISO 11263 (1994). Soil quality: determination of phosphorus – spectrometric determination of phosphorus soluble in sodium hydrogen carbonate solution.

22. Khodaweisi, J., Benejad, H., Afkhami, A., Olyaie, E., Lashgari, S., & Dashti, R. (2011). Synthesisof calcium peroxide nanoparticles as an innovative reagent for in situ chemical oxidation. *Journal of Hazardous Materials, 192*(3), 1437–1440.

23. Liu, X., Song, X., Zhang, Z., Wang, M., & Pan, B. (2014). Non-hydroxyl radical mediated photochemical processes for dye degradation. *Physical Chemistry Chemical Physics, 16*, 7571–7577.

24. Małachowska-Jutsz, A., Miksch, K. (2010). Usefulness of selected bio-indicators to determinate the effectiveness of bioremediation of hydrocarbon-contaminated soil (in polish). PAN, vol. 54, Gliwice.

25. Małachowska-Jutsz, A., Gumińska, M., & Bernaś, Z. (2014a). Effect of calcium peroxideon phytotoxicity of soil contaminated with fluoranthene. *Ochrona Środowiska, 36*(3), 37–42.

26. Małachowska-Jutsz, A., Turek-Szytow, J., & Miksch, K. (2014b). Effect of calcium peroxideon zootoxicity in fluoranthene- contaminated soil. *Przemysł Chemiczny, 93*(12), 2197–2200.

27. Maliszewska-Kordybach, B. (1990). Effect of sandy soil moisture level on the range and rateof fluorene, anthracene and pyrene degradation. *Roczniki Gleboznawcze, 41*(1–2), 47–57.

28. Ostrowska, A., Gawliński, S., Szczubiałka, Z. (1991). Methods of analysis and evaluation of characteristics soils and plants (in polish). Warszawa.

29. Patel, H., Ak, G., & Gupte, S. (2009). Biodegradation of fluoranthene by basidomysetes fungal isolate*Pleurotus Ostreatus* HP-1. *Applied Microbiology and Biotechnology, 157*, 367–376.

30. Patent WO 1994029242 A1 (1994). The bioremediation of soils containing hydrocarbon using calcium peroxide.

31. Pawar, R. (2012). The effect of soil pH on degradation of polycyclic aromatic hydrocarbons (PAHs). PhD thesis, School of Life Sciences, University of Hertfordshire.

32. PN-R-04032. (1998). Soils and mineral soil materials. Soil sampling and determination of particle size distribution in mineral soil material (in Polish).

33. Quasenian, L., Guiral, D., Belghazi, M., Ferre, E., Gros, R., & Farnet, A. M. (2011). Identyficationof various laccases induced by anthracene and contribution to its degradation in a Mediterranenan coastal pine litter.*Chemosphere, 84*, 1321–1328.

34. Quasenian, L., Guiral, D., Ziarelli, F., Ruaudel, F., & Farnet, A. M. (2012). Does anthracene affect microbial activities and organic matter decomposition? A comparative study in *Pinus halepensis* litters from Medirerranean coastal and inland areas. *Chemosphere, 89*, 548–555.

35. Sepic, E., Bricelj, M., & Leskovsek, H. (2003). Toxicity of fluoranthene and its biodegradation metabolites to aquatic organisms. *Chemosphere, 52*, 1125–1133.

36. Sinsabaugh, R. L. (2010). Phenol oxidase, peroxidise and organic matter dynamics of soil. *Soil Biology and Biochemistry, 42*, 391–404.

37. Yang, Y., Shu, L., Wang, X. L., Xing, B. S., & Tao, S. (2012). Mechanisms regulating bioavailability of phenanthrene sorbed on humic substances extracted from a peat soil. *Environmental Toxicology & Chemistry, 31*, 1431–1437.

Chapter 8

Heavy Metal Contamination in Brazilian Agricultural Soils due to Application of Fertilizers

Affonso Celso Gonçalves Jr.[1], Herbert Nacke[2], Daniel Schwantes[3] and Gustavo Ferreira Coelho[4]

[1] Western Paraná State University, Center of Agricultural Sciences, Marechal Cândido Rondon, Brazil

[2] University Center Dynamic of Cataracts, Technological Center, Foz do Iguaçu, Brazil

[3] Pontifical Catholic University of Paraná, Polytechnic School, Toledo, Brazil

[4] Federal University of Paraná, Palotina, Brazil

1. INTRODUCTION

The Brazilian grain production reaches record productivities each year and among the major crops, stand out the maize (*Zea mays* L.) and soybean (*Glycine max* L.). For these two cultures, the obtaining of the maximum yield has a fundamental importance in the current global scenario [1]. In this relentless pursuit of higher productivities, farmers use technologies related to various areas, such as new forms of fertilizer and pesticide application, different fertilizers, crop breeding, equipment and techniques of planting and harvesting. In order to increase the efficiency in this activity, due to the increased requirement for competitiveness of economic globalization, one of the tools found by farmers is the use of fertilizers containing micronutrients in their crops [2]. Among the various factors of production, the need of balanced and sustainable fertilizers occupies a prominent place, which should include not only the primary and secondary macronutrients, but also micronutrients, which are not considered in routine fertilization by most farmers. Currently, with the advancement of the industries of fertilizers, micronutrients are added directly in the formulation of basic fertilizers ($N:P_2O_5:K_2O$), which can be found in the market of domestic and imported agricultural inputs [3]. Among the several micronutrients used in Brazilian and Global agriculture, zinc (Zn) is a major one, having its need scientifically established over 70 years being its main functions related to photosynthesis, respiration, protein synthesis and plant membrane permeability [2].

Nutrient uptake by crops is influenced by various factors, including climatic conditions such as rain and temperature, genetic differences between varieties, type of soil, nutrient content of soil and various cultural practices.

In the strategy of reducing costs, both in manufacture of fertilizers, as well in croplands, the use of industrial waste for the obtaining of micronutrients has been widely used. These wastes, when used appropriately and rationally present themselves as a good solution for agriculture, but their improper use can cause serious damage to soil, plants and human beings. Several researches are touting in scientific works, facts that generate severe criticism to the fertilizer industry, mainly because the involvement of these industries in operations that endanger the environment, where the main factor is the use of raw materials containing toxic metals such as cadmium (Cd), lead (Pb) and chromium (Cr) in the manufacture of fertilizers [4]. In that way, the fertilizers used to supply micronutrients have often, in their composition, in addition to the desirable elements, toxic metals such as Cd, Pb and Cr [5,6].

The purpose of this chapter is to assist in developing a more adequate environmental legislation, that ensures a sustainable environment and a production of quality food. In that way, this chapter highlight the lack of criteria, in the current Brazilian environmental legislation, regarding the allowed limits of toxic heavy metals in agricultural soils and in the use of fertilizers contaminated with heavy metals in agriculture, as well, demonstrate that the use of these contaminated inputs can cause degradation of agricultural soils, making food production unviable at these locations. Furthermore, it was made the study of three cases about the existence of fertilizers contaminated with toxic metals in Brazilian agriculture and the impact caused by its use. The chapter also presents some strategies for recuperation and remediation of soils contaminated with toxic heavy metals.

The presence of heavy metals in fertilizers is a extremely important subject, since investigations in Brazil, coordinated by prosecutors in São Paulo, show evidence of entry loads containing toxic waste through irregular importation and smuggling, several times in Brazilian ports. These investigations shown that companies and manufacturers of agricultural inputs, are importing these toxic chemicals from United States, Canada, Mexico, Spain, Holland and England to be used as raw material in the manufacture of fertilizers to supply micronutrients.

Brazil presents itself as one of most promising nations of contemporary and future world, however, for the effective consolidation of its presence among the great powers is extremely necessary maintain the sustainability of its agroecosystem. Therefore, the Brazilian agricultural activity should be developed without damaging the environment and thus continue to grow in a rational and appropriate way, avoiding harm to future generations. In the world of agribusiness, which needs to be highly competitive, concepts such as traceability, environment and risk assessment are fundamental starting points for those who want to establish and grow.

2. POTENTIAL CONTAMINATION OF SOIL BY AGRICULTURAL ACTIVITIES

Due to its continental dimensions, Brazil has a great diversity of soil types in different regions of the country. Most of these soils have limitations for agriculture, characteristic of tropical regions such as acid, deep and highly weathered soils, with low nutrient availability and high exchangeable acidity (Al^{3+}), resulting in a low cation exchange capacity and low availability of P [7].

These factors highlight the need for adopting management strategies that aim to increase the chances of good crop of the culture when grown in soils of low fertility, giving them the ability to meet the nutritional requirements of these plants. The practices of fertilization and soil amendment aim to make soil suitable for the development of major crops, providing the necessary nutrients for the plants and decreasing the acidity of the soil. These practices are adopted throughout the world, and are based on the prerogative of the need to produce food to sustain the world population that grows rapidly each year.

According to the Food and Agriculture Organization of the United Nations (FAO) [8], in 1980 there were 4.4 billion people on the planet, 5.7 billion in 1995 and 6.9 billion in 2010. Studies estimate that by 2050 there will be 9.4 billion people and in 2200 about 11 billion [9]. Based on this scenario, agricultural activities have expanded greatly since 1945, currently holding approximately 38% of the land surface. Between 1985 and 2005, the acreage increased about 7%, generating large increases in food production [10].

This pressure on agricultural activities promotes an increasing fertilizer consumption per each year that goes around the globe (Figure 1). Studies show that in the early 60's used to be about 146 million tons of fertilizer, increasing in 2010 to 176 million of tons, demonstrating a significant increase in the consumption of these inputs [11]. Also according to the International Fertilizer Association (IFA) [11], Brazil is the fourth largest consumer of mineral fertilizers ($N:P_2O_5:K_2O$) in the world, only behind China, India and the United States. In recent years the consumption of nutrients by Brazilian agriculture has increased considerably being that in 1961 Brazil was only the 25th world consumer, however in the 1990, the country was already appearing as the seventh in the rankings. Regarding to micronutrients, there is an increase in consumption of 13.3 times in the period 1990-2003 [12].

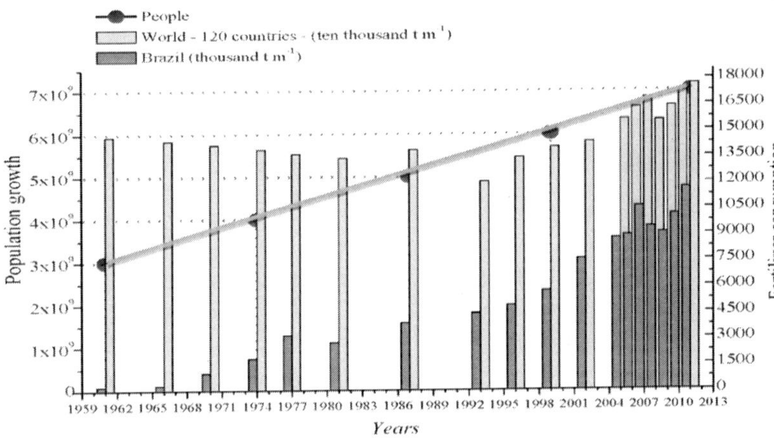

Figure 1. Population growth and consumption of fertilizers in Brazil and in the world [11]

However the use of these products is often not done correctly. The plan for fertilizer recommendation involves planning for the collection of soil samples, soil chemical analysis and recommendation by an appropriate professional, something that many producers do not perform, often using pre-determined recommendations without professional assistance. Another problem is the lack of crop rotation and the use of highly productive cultivar which, according to reference [13], require intense macronutrient application, resulting in micronutrient deficiencies in the soil.

Overuse of fertilizers can also provide soil contamination, it increases the concentration of a particular element or substance of environmental interest above its naturally occurring in the soil [7]. This contamination is related to eutrophication (excess nutrients), in this case the major nutrients to be considered are nitrogen (N) and phosphorus (P). Another concern is the availability of harmful elements such as toxic metals. These elements can express their pollutant potential directly on soil organisms, due to the availability to plants in phytotoxic levels, plus the ability to transfer to the food chain through the own plants, or by contamination of soil and water resources [14].

The presence of harmful elements in fertilizers is worrying because until the 70's, most fertilizers were obtained directly from ores found in nature, and although they also contained toxic metals, the beneficiation process of this material was less offensive from the environmental standpoint, because the contaminant levels were very low [6]. Recently, the shortage of mineral resources and modern extraction methods required large investments in equipment and technology, creating a technical difficulty and the pursuit of lower costs of production, which eventually encouraged companies to seek a cheaper solution such as using industrial waste, that may be feasible since it does not contaminate the environment. However, according to the Environmental Sanitation Technology Company of São Paulo State (CETESB) [15], in São Paulo, until the end of 2012, 4572 contaminated areas were found, and approximately 4% of these areas were contaminated with industrial waste, and among the contaminants, toxic metals are in 4th place, behind Polycyclic aromatic hydro carbons contaminants (PAHs), aromatic solvents and liquid fuels.

Between the sources of soil contamination from waste, there is the contamination of sewage sludge, that even though has a variable composition, with macro, micro nutrients and organic matter, their use may be limited by the presence of pathogens and toxic metals [16,17]. Studies evaluating the use of sewage sludge in sugarcane found that the doses used increased levels of chromium, nickel (Ni), Pb and zinc (Zn) in the soil [18].

Reference [19] studying the availability of Cd in dystrophic soil submitted to the application of sewage sludge, observed that at the first hour of contact between the sewage sludge and soil occurred transference of Cd to the solution. After this interval, the Cd concentration in the solution decreases, occurring stabilization in the following hours, due to adsorption and ion exchange which act providing less metal. According to the authors, in all observed time intervals, the levels of Cd were higher than reference values for CETESB (<0.5 mg kg^{-1}) [20], which characterize the contamination of the soil by sewage sludge.

The use of pig manure as fertilizer is also a practical often carried out mainly in south of Brazil, however, some studies have shown risks as to its overuse. Reference [21] evaluated the accumulation of metals in the soil under the systematic application of swine manure, and found an increase in the availability of copper (Cu), zinc and manganese (Mn). The authors also claim that in these cases there is a need to monitor these elements over time, so that the total contents do not exceed the critical values. In study conducted in China [22], the application of manure in soil for 16 years resulted in higher accumulation of Cd. One of the major sources of metals such as Pb, Ni, Cr and Cd in pig manure is mineral supplied by commercial feed [21].

The metals are also components of many pesticides, mainly Cu, Zn and Pb, which cause high soil contamination by these elements [23], another case is the use of Cu, that has been used since 1917 for the Bordeaux mixture [24]. In a study conducted in Caetés watershed in the county of Paty's Ensign, Rio de Janeiro, Brazil, it was aimed to evaluate the soil contamination by toxic metals, from the intensive use of agrochemicals, in which the authors found an increase in the total levels of metals in soil, stream water and dams. The soil metal concentration not reached critical established levels, however the result was worrying about the water stream and dam, which showed total contents of Cd, Mn and Pb above maximum standards established by the Brazilian legislation [23].

In England and Wales, reference [25] developed an inventory on the input of metals in agricultural soils. Forms of contamination are the most diverse, the main sources are generated by atmospheric deposition, sewage sludge, manure, inorganic fertilizers, lime, agrochemicals, irrigation water and industrial waste.

All these reports point to industrialization as the main cause for the large amount of metal dispersed in the environment, and human as the protagonist of all this pollution, because usually mankind seeks the easiest way supply their needs while generating countless volumes of waste. These actions put us in the position of predator. And thus there is a need of legislations that seek to limit the presence of these contaminants in the soil. In Table 1, the international maximum levels of metals allowed in soils of various places show what would be, in theory at least, a concern of many countries in the world about metal contamination.

Table 1. Maximum permitted levels of toxic metals in the soil in some countries

Countries	Toxic Metals (mg kg^{-1})				
	As	Cd	Pb	Cr	Hg
(*)European Comunities	-	1-3	50-300	-	1-1,5
(**)United States	14	1.6	50-300	1500	0.5
(**)Poland (1977 – 1993)	30	1-3	70-150	50-80	5
(**)Germany (1984 – 1992)	20	3 – 1.5	100	100	2
(**)United Kingdon	10	3-15	500-2000	-	-
(**)Austria	50	5	100	100	5

Source: * [26], ** [27]

Currently in Brazil, the legislation that deals about soil contamination is Resolution n°. 420 of 2009 of the National Council of Environment (CONAMA) [28], which aims to establish the criteria and values that guide of soil quality as the presence of chemical substances, among them the toxic metals. The resolution establishes three categories of guiding values which are: The Reference Value Quality (RVQ) which is the concentration of a substance that will define the natural quality of the soil, based on previous studies for each Brazilian state, the Prevention Value (PV), which refers to the concentration limit of a substance in the soil, and the Investigation Value (IV) which is the concentration of a substance in the soil of which there are potential risks, direct or indirect, to human health. Table 2 shows the values set for preventing and detecting toxic metals in soil.

However, the Resolution n°. 420 is being severely questioned by authorities and researchers, since it present several misconceptions that violate the Federal Constitution regarding the "preservation, enhancement and restoration of environmental quality propitious to life" (this issue will be addressed with greater emphasis on section 5 of this chapter).

3. FERTILIZERS *VERSUS* CONTAMINANTS

There are three different categories of agricultural fertilizer: macronutrient fertilizer to supply major nutrients (N, P and K); macronutrient fertilizer to supply secondary nutrients [calcium (Ca), magnesium (Mg) and sulfur (S)] and micronutrients [Cu, Zn, Ni, Mn, iron (Fe), molybdenum (Mo), boron (B), chlorine (Cl), sodium (Na), cobalt (Co) and silicon (Si)]. All of these nutrients can be applied through simple fertilizer (only one nutrient) or compound form (formulations with more than one nutrient).There is a further classification based on the essentiality of nutrients, where metals may be classified as essentials, such as B, Cu, Fe, Mn, Mo and Zn; beneficial, such as Co, Ni and V; and toxic, such as Cd, Cr, Hg and Pb. Thus, it should be noted that the essential and beneficial elements can become toxic at high concentrations [30].

Table 2. Values guiding to the presence of toxic metals in Brazilian soils in accordance with Resolution n° 420 of 2009 of CONAMA [28]

Metal	PV	IV		
		Agricultural Area	Residential	Industrial
As (mg kg^{-1})	15.0	35.0	55.0	150.0
Cd (mg kg^{-1})	1.3	3.0	8.0	20.0
Pb (mg kg^{-1})	72.0	180.0	300.0	900.0
Cr (mg kg^{-1})	75.0	150.0	300.0	400.0
Hg (mg kg^{-1})	0.5	12.0	36.0	70.0

PV: Prevention Value; IV: Investigation value

According to reports developed by the United States Environmental Protection Agency (USEPA) [31] and The Weinberg Group (requested by The Fertilizer Institute) [32], the main metal contaminants that are found in agricultural fertilizers are Cd, Cr, Co, Cu, Pb, Mo, Ni, Zn, arsenic (As), mercury (Hg) selenium (Se), vanadium (V) and radio 266 (Ra). It can be observed in this list, that among the metals commonly found as contaminants in fertilizers, occur the presence of micronutrients, required for plant development.

According to USEPA [31], among the fertilizers used to supply macronutrients, the phosphate shown the highest levels of toxic metals (especially Cd), followed by nitrogen and potassium fertilizers. With respect to micronutrients, the sources used to obtain it are varied, resulting in the occurrence of fertilizers from low to high levels of metals. In general, micronutrient sources have higher level of contaminant in comparison to sources of macronutrients, on the other hand, are applied larger amounts of fertilizer to supply the macronutrients, therefore, the possibility of soil and environment pollution with the use of any type of contaminated fertilizer is real and should be seriously studied.

The presence of contaminants in fertilizers occurs most of the time with the use of industrial waste or low cost sources for raw materials. These sources, when treated and used properly and rationally, present themselves as a good alternative to agriculture, however its improper use can cause serious damage to the soil, plants and the human beings.

This fact becomes even more important in the case of fertilizers to supply micronutrients, which often are obtained from raw materials originated from industrial wastes, which may show, in addition of the desirable elements, toxic metals. In this situation, the benefits are highly attractive for the fertilizer industries, once these companies acquire low quality raw material at low costs, without generating any type of waste, incorporating everything in their products.

One of the main cases regarding the presence of contaminants in fertilizers happened in the 90s in the United States, where many farmers and social actors reported the possibility of the presence of toxic metals in these inputs. The impact generated and population pressure led to the development of several research regarding the contamination of fertilizers and soil pollution, culminating with the publication by the U.S. government, of the first law regulating the content of toxic elements in the United States in 1998. From these studies and based on Canadian law, in 2002, the USEPA published the federal law that regulates the production of fertilizers to supply Zn from industrial waste [33].

Another case reported in the literature is the presence of contaminants in raw materials from China. In a publication from the Washington Department of Ecology (WDE) [34] it was reported the presence of Cd in raw materials for the manufacture of fertilizers, with samples showing more than 20% of Cd, i.e., more than 200,000 ppm. It is also reported that the final products produced with the use of these contaminated sources presented 10,600 ppm of Cd. The report also warns that Cd present in the analyzed material does not naturally occur in natural mineral sources, leading to the conclusion that the incorporation of Cd or the use of wastes contaminated with this metal was occurred.

In Brazil, there are frequent reports of companies selling fertilizer, mineral and organic, outside the standard quality. The most common cases are related to the presence of nutrients amounts below the specified labels, since this factor is more investigated. However, there are few studies evaluating the presence of toxic metals in these inputs, although the few that were conducted demonstrated the presence of these contaminants in fertilizers [6, 35, 36].

Currently, most grain-producing countries and other plant materials have laws regarding the limits of metals in fertilizer, which are presented in Table 3, where it can be observed a large variation between the allowed values. In general, the stipulated limits are for metals in phosphate fertilizer and micronutrient fertilizers, but there are some normatives that consider fertilizers in general, regardless of its composition or purpose. The limits of toxic metals can also be determined based on the amounts of nutrients or total amount in fertilizers.

Table 3. International maximum limits of contaminants allowable in fertilizers

Switzerland[1]	Phosphate	NA	50.0	NA	NA
Finland[1]	Phosphate	NA	50.0	NA	NA
Sweden[1]	Phosphate	NA	100.0	NA	NA
Norway[1]	Phosphate	NA	100.0	NA	NA
Denmark[1]	Phosphate	NA	110.0	NA	NA
Belgium[1]	Phosphate	NA	210.0	NA	NA
Germany[1]	Phosphate	NA	210.0	NA	NA
Austria[1]	Phosphate	NA	275.0	NA	NA
U.S. (California)[2]	Phosphate	2.0*	4.0*	20.0*	NA
U.S. (Washington)[3]	Phosphate	13.0	165.0	61.0	1.0
Australia[1]	Phosphate	NA	300.0	NA	NA
Japan[1]	Phosphate	50.0	343.0	100.0	5.0
New Zealand[1]	Phosphate	NA	280.0	NA	NA
U.S. (Texas)[4]	Anyone	41.0	39.0	300.0	17.0
Canada[1]	Anyone	75.0	20.0	500.0	NA
China[4]	Anyone	50.0	8.0	100.0	NA
U.S. (California)	Micronutrient	13.0**	12.0**	140.0**	NA
U.S. (Washington)	Micronutrient	112.0**	83.0**	463.0**	6.0**
Australia	Micronutrient	NA	50.0	2000.0	5.0

1 [38], 2 [39], 3 [40], 4 [41]; * - Maximum allowed value per percentage point of P_2O_5 in the fertilizer; ** - Maximum allowed value per percentage point of the micronutrients sum in the fertilizer ; NA – not available

The Brazilian fertilizers law is based on the Normative Instruction (NI) 27/06 [37], which states like contaminating toxic metals, the elements As, Cd, Pb, Cr and Hg, determining the maximum allowable limits of these metals in four different categories of fertilizers (Table 4).

There are some doubts in relation to NI 27/06, which consider the legislation incipient, since that it does not determines which raw materials can be used as sources of nutrients in the fertilizers. Nutrient sources for macronutrients generally present good quality, however, for micronutrients, the use of low quality materials is frequently observed in Brazil, as mentioned previously, resulting in a possible contamination of the environment.

Table 4. Maximum allowable limits for metal contaminants in Brazilian fertilizers [37]

Fertilizer type	Maximum permitted quantity of the metal contaminant (mg kg^{-1})				
	As	Cd	Pb	Cr	Hg
Simple mineral fertilizer to suply P_2O_5	2.0*	4.0*	20.0*	40.0*	0.05*
Compound mineral fertilizer with micronutrient	500.0**	15.0**	750.0**	500.0**	10.0**
Compound mineral fertilizer with macronutrient and micronutrient	250.0	57.0	1000.0	NA	NA
Mineral fertilizer to suply micronutrient	4000.0	450.0	10000.0	NA	NA

* - Maximum allowed value per percentage point of P_2O_5 in the fertilizer; ** - Maximum allowed value per percentage point of the micronutrients sum in the fertilizer; NA – Not Available.

The legislation also does not tell how it will be the inspection of fertilizers industries, moreover tolerating the presence of contaminants in amounts up to 30% higher than the accepted values (Table 4). Another worth mentioning fact, is that the limits should be revised, since Article 5 affirms the need of a revision of the values predicted by NI 27/06 four years after its publication.

Nowadays, still does not exist an adequate legislation in Brazil regarding the use of toxic waste in the production of micronutrients. However, there are technical studies which classify these elements as hazardous materials, an example is the NBR 10004 of the Brazilian Association of Technical Standards (ABNT) [42], which stablish that the use of this wastes may present a risk to public health, causing or contributing to an higher incidence of diseases and mortality, also presenting adverse effects on the environment when handled or disposed inappropriately.

Another issue that is being discussed by the scientific community is that the determination of contaminants limits in fertilizers provides for the industries the possibility to include, as a source of nutrients, contaminated materials or with low quality, i.e., by the laws, the government provided the industries, exactly what they wanted, the possibility to incorporate low-quality raw materials and contaminated to supply nutrients [43]. It is observed so that there is a strong influence of the fertilizer industry in the studies and laws that limit the contaminants in these materials, showing that this problem is not only related to the environmental field, but also with the political and economic sector.

4. METALS: TOXICITY IN THE FOODCHAIN

Due to its mobility in aquatic systems and to its toxicity to life forms, the toxic metals or heavy metals, once in surface or in groundwater, are mentioned as one of the most problematic inorganic contaminants in the environment.

The toxic meals are strongly pollutants elements and present many harmful effects to ecosystems, causing physical-chemical alterations to water, what promote the decrease of water quality and mortality of the fauna and flora, prejudicing consequently, the human health [44].

The term heavy metal, is many times used with connotations of pollution and toxicity. However, "heavy" is conventionally used to imply the high density of the metal, referring to the pure element or to an alloy of metallic elements. The expression heavy metal apply to elements which have specific mass higher than 5 g cm^{-3} or which have atomic number higher than 20 [6].

Some metals, as the case of Cd, Pb, Ni and others, even when its presence is diluted in waters, many times in undetectable quantities, have recalcitrant effect, i.e., its persistence in the environment imply in process as biomagnification, which regards to the accumulation of these harmful substances in the many trophic levels of the ecological chains of ecosystems.

The contamination by toxic metals in animals can occur due to inhalation, skin penetration or ingestion of contaminated waters and foods, this last caused mainly by the capacity of plants in bioccumulate these elements in high proportions and later offer them to animals and human beings [45].

In animals, the toxic metals act in highly specific reactions, enzymatic in the most part and in alterations in systems that evolves these reactions, resulting in negative effects. One classic example is the metalloenzymes, in which dimensional structure is changed if one "intruder metal" binds to its active site, what prevents the enzymatic action [46].

Metal contaminations in the environment results from a sum of natural processes, between which also make part the natural sources of contamination, however, it is a common sense that the sources so called anthropogenic are more frequent when compared to the natural sources of contamination [47].

Significant increases of the metals concentrations are found in areas of intense industrial activity, where the accumulation can be many times higher than the average levels from non-contaminated areas. However, areas distant from the industrial centers in many cases also show metals contamination, due to the atmospheric transportation of these compounds [48] and in the case of agricultural soils, due to the use of fertilizes contaminated with these metals [6].

The contamination by metals is arising from the continuous disposal of industrial, domestic or livestock wastes in soils, rivers, lakes and seas. These pollutants include the toxic metals, which cause great concern to public health, because they are incorporated very easily in the tropic chain. These harmful elements cannot be destroyed by biological degradation, as in the case of the most part of organic pollutants. Incidences of the accumulation of metals in fish, oysters, mussels, sediment and other components from the aquatic systems have been reported in many countries around the world [49, 50].

Once the metals are present in cultivated soils, they are likely of radicular absorption by plants. It is important to point out that some of these elements present biological function whereas others are much known as toxic even in low concentrations. This way, the plants can be considered as constituents of the foundation of the trophic chain, causing great concern due to the possibility of the occurrence of toxic concentrations of certain metals which are transported from plants to higher levels of the food chain [51].

The toxic metals are not only subjected to radicular absorption by plants when presents in the soil solution, because may also occur the fixation of the metals in the soil particles, seepage/leaching to lower levels of soil or even the surface runoff, reaching the groundwater, rivers, lakes and seas. In general, there are two types of soils in Brazil, sandy and clayey, so that in sandy soils, due to the low ion exchange capacity, the metals are easily moved in the environment by leaching or surface runoff, reaching easily the hydric resources. In the other hand, in clayey soils, due to the higher ion exchange capacity, the metals tend to stand immobilized in the soil particle, becoming available for the plants in more than one cultivation season.

The toxic metals resulting from the anthropogenic activities, when disposed incorrectly in the environment, directly or indirectly reaches the top of the chain, i.e., humans will have contact with these elements by the consumption of food from vegetal or animal origin, contaminated by water, soil or air. Once in the human organism, these metals present cumulative character, i.e., are accumulated in the live tissues, i.e., hardly are eliminated by the organism. Down below are related some of the characteristics of some toxic metals and its respective collateral effects caused to human organisms.

4.1. Arsenic (As) – Cas Id 7440-38-2

In humans, the ingestion of high levels of arsenic can result in death. Exposure to lower levels can cause nausea and vomiting, decreased production of red and white blood cells, abnormal heart rhythm, damage to blood vessels, and a sensation of "pins and needles" in hands and feet. The ingestion of inorganic arsenic can increase the risk of skin cancer and cancer in the liver, bladder, and lungs. Inhalation of inorganic arsenic can cause increased risk of lung cancer. Exposure to higher than average levels of arsenic occur mostly in the workplace, near hazardous waste sites, or in areas with high natural levels. At high levels, inorganic arsenic can cause death [52].

Aquatic organisms accumulate arsenic mainly as inorganic forms, and some of the organisms such as phytoplankton, bacteria, etc. transform them into methylated and organic forms [53].

As is associated with ores containing metals, such as copper and lead. Arsenic may enter the environment during the mining and smelting of these ores. Small amounts of arsenic also may be released into the atmosphere from coal-fired power plants and incinerators because coal and waste products often contain some arsenic. It cannot be destroyed in the environment. It can only change its form, or become attached to or separated from particles. It may change its form by reacting with oxygen or other molecules present in air, water, or soil, or by the action of bacteria that live in soil or sediment [52].

Researching the effects of arsenic accumulation from plants cultivated near pressure-treated wood, Cao and Ma [54] found out that elevated As concentrations were observed in the soils adjacent to fences and poles, because As was frequently used for wood treatment, in order to increase the wood life time. Eating vegetables from these contaminated soils may pose a risk of As exposure since vegetables would readily take up As from these soils.

As, according to the Agency for Toxic Substances and Disease Registry (ATSDR) [55] is the first substance on the Priority List of Hazardous Substances.

4.2. Cadmium (Cd) – Cas Id 7440-43-9

At room temperature (25 °C/73.4 °F) the Cd is presented at solid state, with atomic number of 48 and atomic mass of 112.41 g. This element belong to the group of the metal of gray silver metallic color and it is available in many forms including sheet, granule, pellet, powder, rod, wire, among other. It is a soft metal, highly toxic which can be easily cut with a knife, and mostly produced as a byproduct of the extraction of Zn, Cu and Pb [56].

It is considered one of the most toxic metals, presenting important harmful effects for biological activity on soil, plant metabolism, human and animal health. The abundance of Cd in magmatic and sedimentary rocks, generally not more than 0.3 mg kg^{-1}, being that this metal tends to be concentrated in clayey deposits. The Cd is strongly associated to Zn in the geochemistry, however, have a strong affinity with S, presenting a higher mobility than Zn in acid environments [27].

Generally ore (CdS) is associated as a contaminant (0.25 to 0.5%) to ore Zn and Pb. According to the International Cadmium Association [57], Cd has been mainly used in manufacturing NI-Cd batteries, though, also has other applications as pigments, coatings, stabilizers for PVC (polyvinyl chloride) and alloys.

One of the main problems associated with Cd is its final destination, because it can reach the soil or the air through the burning of urban waste or the burning of fossil fuels, in this way polluting the environment and causing damage to the ecosystem [58].

In humans, Cd affects several organ systems, as the Cardiovascular (Heart and Blood Vessels), Developmental (effects during periods when organs are developing), Gastrointestinal (Digestive), Neurological (Nervous System), Renal (Urinary System or Kidneys), Reproductive and Respiratory (From the Nose to the Lungs), Cd is also known as a carcinogen agent [52]. According to the Agency for Toxic Substances and Disease Registry [55] Cd is the seventh substance on the Priority List of Hazardous Substances.

4.3. Lead (Pb) – Cas Id 7439-92-1

The Pb is rarely found in its natural state, but in combination with other elements, being the most important ores which present this metal are: Galena, cerussite, anglesite, perite and others.

Among the toxic metals, Pb is present in larger amounts in the earth's crust, with an average concentration between 10 and 20 mg kg^{-1} [59], therefore, is among the most used metals worldwide scale of Fe> Al> Cu> Zn> Pb [60].

To be quite versatile, Pb is much used in industrial products, and can be hardened by the addition of other metals and metal alloy being used as producing components, welds and other materials, in addition to being able to form various compounds employed in chemical industry [61].

According to unofficial information from the National Department of Mineral Production (DNPM), the main consuming sectors Pb are: accumulators (batteries) 80%, representing approximately 88 tons of metal, oxides, 12%, equivalent to 13 tons; and electronic (alloys, solders and various), 8%, with 9000 tons.

The Pb in soil is extremely stable and highly toxic to humans and animals, is currently ranked as the second most dangerous element in the Priority List of Hazardous Substances [55]. Most of the lead enters the human body through the respiratory and gastrointestinal routes, and after absorption, it can be found in the blood, soft and mineralized tissues [62].

In humans, the Pb poisoning may initially cause lack of appetite, metallic taste in mouth, muscle discomfort, malaise, headache and abdominal cramps strong. However, in childhood, symptoms often linked to the deposition of this metal in the brain are predominant. In its inorganic Pb^{2+} is an inhibitor of enzymes, further affecting the nervous system [63].

The organ systems affected by Pb are the Cardiovascular (Heart and Blood Vessels), Developmental (effects during periods when organs are developing), Gastrointestinal (Digestive), Hematological (Blood Forming), Musculoskeletal (Muscles and Skeleton), Neurological (Nervous System), Ocular (Eyes), Renal (Urinary System or Kidneys) and Reproductive [52].

4.4. Chromium (Cr) – Cas Id 7440-47-3

With regard to heavy metal Cr, this has turned into a serious pollutant of air, soil and water. About 70% of the production of Cr is used in alloys, including stainless steel and 15% in industrial chemical processes, mainly in leather tanning, pigments, electrochemical and cooling towers [64].

The ions of Cr are found in different oxidation states, however, the trivalent form (Cr^{3+}) is the most common, while the hexavalent form (Cr^{6+}) depends on the more specific conditions of pH and redox potential [65].

The toxicity of Cr depends on its oxidation state. While the Cr^{3+} is relatively stable and immobile, the Cr^{6+} moves easily through the soil and aquatic environments, with a strong oxidizing agent capable of being absorbed even by the skin in this way, the Cr presents itself as a carcinogenic species [66, 67].

In humans, the Cr affects the immune, renal (kidney or urinary system) and respiratory tract (the nose to the lungs), and is known as a powerful carcinogen [52].

According to reference [68], a lethal dose of Cr^{6+} is between 50 and 100 mg kg^{-1}, a value much lower when compared to Cr^{3+}, the values are between 1900 and 3300 mg kg^{-1}. Since the Cr^{6+} occupies the 17th position in the Priority List of Hazardous Substances [55].

4.5. Nickel (Ni) – Cas Id 7440-02-0

The Ni occurs in the earth's crust in the order of 0.01%, mainly as sulfide, oxide and silicate minerals. Geological activities as natural weathering and volcanoes led the natural distribution of this metal in natural environments at modest levels. The Ni metal and its compounds are used in multiple applications as stainless steel and other alloys, castings, catalysts, batteries, electronics, ceramics, pigments and even coins [69].

Among the main uses of this metal can still cite it use in coating parts, in a process called electroplating, widely used around the world by Electroplating industries [70].

Although not an element extensively released to the environment, this can present a risk to human health. Epidemiological investigations and experimental studies have shown that certain compounds of Ni are extremely carcinogenic after inhalation, however, the same studies show that these risks are limited by conditions specific occupational exposure to this metal [71].

Results show that this element can cross the placenta and present embryological and teratogenic properties. The main danger of Ni to the man beside the carcinogenicity is related to the ability to cause sensitivity reactions [71].

The most common harmful health effect in the general population is allergic contact dermatitis elicited by prolonged skin contact of sensitized individuals with Ni. Therefore, the majority of studies related to health effects of Ni to humans concern airborne Ni particulate matter. Strongest evidence for high cancer risk was found for sulfidic nickel species (NiS, NiS_2 and Ni_2S_3) in Ni refinery dust. On the molecular level the toxic Ni species responsible for severe health effects, as allergic contact dermatitis and respiratory tract cancer, has been suggested to be caused by Ni^{2+} [69].

Ni affects the following organ systems: Cardiovascular (Heart and Blood Vessels), Dermal (Skin), Immunological (Immune System), Respiratory (From the Nose to the Lungs) [52]. Actually, Ni ranks the 57th position on the Priority List of Hazardous Substances [55].

4.6. Copper (Cu)- Cas Id 7440-50-8

The Cu, despite being a heavy metal, meets the criteria of essentiality for plants and microorganisms, being classified as micronutrient [72]. This metal is essential to human health, and is present in all human tissues and many fluids and with amino acids, fatty acids and vitamins, is required for the metabolic processes [73].

Excess copper may be toxic to humans because of the affinity metal with S-H groups of many proteins and enzymes associated with diseases such as epilepsy, melanoma and rheumatoid arthritis, as well as the loss of taste [72]. Toxic levels of Cu occur naturally in some soils, whereas others may contain high levels of Cu as a result of the anthropogenic release of contaminated materials into the environment through mining, smelting, manufacturing, agriculture and waste disposal technologies [72].

In plants, either deficient or in excess, Cu can cause disorders in plant growth and development by adversely affecting important physiological process. In particular photosynthetic electron transport is altered under deficiency and excess of Cu. Thus, for healthy plant growth and development Cu must be acquired from the soil, transported throughout the plant, distributed and compartmentalized within different tissues and its content carefully regulated within different cells and organelles. For this purpose, plants, like all other organisms, have homeostatic mechanisms to maintain the correct concentrations of essential metal ions. Membrane transport systems are likely to play a central role in these processes [73].

Copper ranks the 125th position on the Priority List of Hazardous Substances [55].

4.7. Manganese (Mn)- Cas Id 007439-96-5

According to the International Manganese Institute (IMnI) [74], this metal is the fourth most used in terms of tonnage, being ranked behind Fe, Al and Cu, being the production of Mn alloys about 17.7 million metric tons at 2011. Mn the second major abundant metal and 12th most abundant element in the earth's crust, being its ores deposits usually of sedimentary origin, with oxide ore layers inter-bedded with iron-rich formations [75].

Steel is basically an alloy of iron and carbon, consisting of an iron phase and iron carbides. Crude steel produced from iron contains an undesirable amount of oxygen and some sulphur. Mn plays a key role because of two important properties: its ability to combine with sulphur and its powerful deoxidation capacity [76].

High grade Mn ores (> 40%) are typically processed into suitable metallic alloy forms by pyrometallurgical processes. Low grade manganese ores (< 40%) are conventionally processed by pyrometallurgical reductive roasting or melting followed by hydrometallurgical processing for production of chemical manganese dioxide (CMD), electrolytic manganese (EM) or electrolytic manganese dioxide (EMD) [77].

In the human body, Mn displays a somewhat unique behavior with regard to its toxicity. It is relatively non-toxic to the adult organism except to the brain where it causes Parkinson-like symptoms when inhaled even at moderate amounts over longer periods of time [78].

This element, when in large amounts, affect fertility in mammals and are toxic to the embryo and fetus, what implicates that pregnant women should not be exposed to Mn anytime [78].

Due to the substitution of Pb by Mn in gasoline, reference [79], researching the cultivation of beans and oats in soil next to roads, found not only Mn concentrations significantly higher in organic soil but also higher accumulation found in fruits and stems, suggesting that the addition of MMT to gasoline can be the cause for this increase in exchangeable Mn in organic soils.

According to the Agency for Toxic Substances and Disease Registry [55], Mn is the 140th substance on the Priority List of Hazardous Substances.

5. CASE STUDIES: CONTAMINATED FERTILIZERS IN BRAZILIAN CROPS

Based on information obtained from seizures made in Brazilian seaports, related to the entry of loads of industrial waste with the presence of toxic metals for use as source of micronutrients in fertilizers [6], three studies were conducted to evaluate the possibility of soil and plants contamination using different sources and Zn levels, since this is main micronutrient applied in the Brazilian agricultural crops. All studies were conducted at the State University of West Paraná - Brazil, being one for soybean [2], one for the wheat crop [4] and the last one for the corn crop [5].

In the work done with the soybean crop [2], an experiment was conducted in 2007, in Palotina-PR, with coordinates 24° 18' S and 53° 55' W (Figure 2) in a soil classified as Hapludox (RED). It were used four different fertilizers for the Zn supply sold in Brazil, in three doses (0.0; 0.9 and 1.8 kg ha^{-1}). The results of this study demonstrated the presence of toxic metals (Cd, Pb and Cr) in the soybean leaf tissue (Table 5) at levels above of the tolerable in agricultural crops [27]. We also observed the occurrence of these three metals in the soil after harvest (Table 6), which compared to the Brazilian legislation, that establishes the maximum concentrations of contaminants in soils [28], classify the obtained values as above the maximum allowed for the soil be agriculturable (IV - Agricultural area) in the case of Cd. For Pb and Cr the values obtained in the study were below of the prevention value (PV), which is the maximum allowable value of the contaminant in soil to that it is able to sustain its biological processes.

The work done with the wheat crop [4] was conducted on the residual fertilization of the work with the soybean crop, since the concentrations of the metals Cd, Pb and Cr in the soil of the experiment were significant, especially in the case of Cd. The results of this study showed no accumulation of Cd in the leaves of wheat at the level of detection by the analytical method (flame atomic absorption spectrometry - FAAS), however, we found concentrations of Pb and Cr. For Pb the maximum concentration found was under the tolerable limit, and for the Cr, the maximum concentration found was above the tolerable level for crops [27]. One of the main conclusions of this work was the occurrence of availability of Cr and Pb to wheat plants with fertilization performed in previous crop (soybean), demonstrating the persistence of contaminants in soil and their occurrence in more than one crop agriculture.

The work conducted with maize crop was conducted in Mercedes, Paraná State-Brazil, with geographic coordinates of 24° 25' S and 54° 15' W (Figure 3) in a Rhodic Eutrudox. The fertilization of this experiment was performed with eight different sources of Zn in four different doses (0.0; 2.0; 4.0 and 6.0

kg ha^{-1}). In the chemical analysis performed in the fertilizers used in the work, it was observed that one of the sources presented 11039.46 mg of Pb kg^{-1}, being the maximum limit permitted by Brazilian law (NI 27/06) [37] 10000.00 mg of Pb kg^{-1} (Table 3). However, as previously mentioned in the section 3, the NI 27/06 tolerates the presence of contaminants up to 30% of the maximum permitted value, in this way, the fertilizer would not be considered as irregular, and can be freely traded.

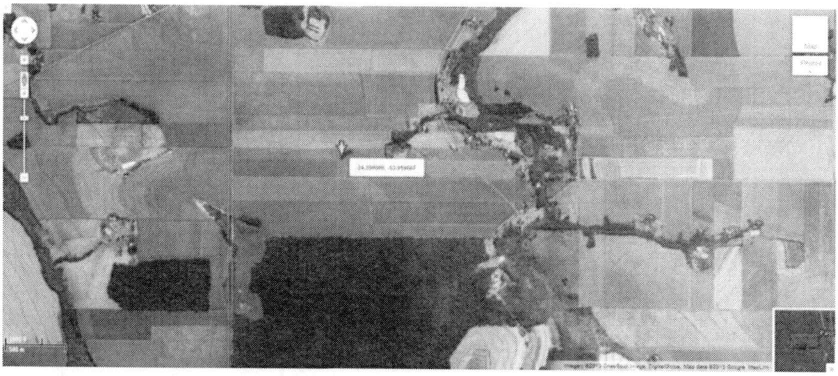

Figure 2. Local where the experiment involving fertilization with Zn in soybean and wheat culture were performed [80]

Figure 3. Local where the experiment involving fertilization with Zn in corn crop was performed [80]

Regarding the leaf concentration of Cd, Pb and Cr (Table 5), it was observed that only the Pb concentrations was found at detectable levels by the used method (FAAS), where the leaf content of Pb in corn plants can be considered inside the tolerable limits [27]. For the soil, it was observed the occurrence of Cd, Pb and Cr (Table 6), so that the concentrations of Cd were below the maximum allowable limit so that the soil is agriculturable (IV - Agricultural area), and the Pb and Cr were below the PV, which is the maximum allowable value of the contaminant in the soil, to that it be able to sustain its biological processes.

Table 5. Foliar levels of Cd, Pb and Cr in experiments conducted with mineral fertilizers to supply Zn in different crops in Brazil

Metal	Agricultural crop			Tolerable foliar level****
	Soybean*	Wheat**	Corn***	
Cd (mg kg^{-1})	0.73	ND	ND	0.05 a 0.50
Pb (mg kg^{-1})	25.53	1.95	8.67	0.50 a 10.00
Cr (mg kg^{-1})	3.36	9.28	ND	2.00

ND – not detected by the method used (FAAS) with a quantification limit of 0.01 mg L^{-1}; * - results from reference [2]; ** - results from reference [4]; *** - results from reference [5]; **** - Tolerable foliar level according to reference [27].

Table 6. Levels of toxic metals Cd, Pb and Cr in soil after cultivation of various crops fertilized with Zn in Brazil

Metal	Agricultural crop		Classification of the metal value in the soil*	
	Soybean*	Corn**	PV	IV – Agricultural area
Cd (mg dm^{-3})	5.52	2.52	1.30	3.00
Pb (mg dm^{-3})	24.25	51.48	72.00	180.00
Cr (mg dm^{-3})	22.19	18.43	75.00	150.00

* - results from reference [2]; ** - results from reference [5]; *** - Classification of the metal value in the soil according to reference [28]; PV–Prevention Value; IV – Investigation Value for agricultural area

Based on the results obtained in the three presented studies it can be inferred that the presence of metal contaminants in fertilizers is real and should be studied more carefully by regulatory authorities in Brazil. The foliar and soil levels of the metals Cd, Pb and Cr found cause concern from the environmental point of view, due the possibility of entry of these contaminants in the food chain, either by consumption of contaminated materials or pollution of water resources from contaminants in the soil. It should be noted that the values presented in these works were obtained with only one fertilization, so that with successive fertilizations may occur a greater availability of these contaminants to the environment.

It is worth affirm also that the laws related to the presence of contaminants [37], and levels of toxic heavy metals allowed in Brazilian soils [28], are being severely questioned by Brazilian researchers and government agencies. Among the key questions are the lack of criteria for determining and monitoring the presence of contaminants in fertilizers, passivity with the addition of toxic metals in brazilians soils by the minimum quality and lack of discussion with researchers and entities responsible for the drafting of legislation.

It should furthermore be highlighted that before approval of Resolution No. 420 of CONAMA [28], there was serious questioning on the part of Brazilian authorities and researchers in relation to the criteria used for its formulation by means of Proceeding 02000.002955/2004-69 [81-83], thus confirming the need for more societal discussion (especially among the specialized technical and scientific segments in a multidisciplinary manner) before this is submitted for approval of possible resolution. The resolution, moreover, violates the Brazilian Federal Constitution with respect to the "preservation, improvement and recovery of the environmental quality favorable to life" because it permits the introduction of contaminants in the soil up to the maximum limit permitted by the resolution, thus allowing the use of contaminated fertilizers and soil conditioners in agriculture until they reach and exceed these values and thereby degrade environmental quality. Therefore, the need for more in-depth field studies in relation to environmental legislation before it is drafted and approved becomes clear. These studies must be undertaken together with environmental regulatory bodies and research entities in the country. Based on the results obtained in this study, it is evident that some of the Zn fertilizer sold by sources in Brazil significantly affect the quantity of Cd, Pb, and Cr in soil. If their use is continued, there may be contamination of soil, surface and underground waters, plants, animals, and humans, leading to severe damage to health as well as the environment. This fact raises concern because according to references [84-86], even low levels of heavy metals in soil may be available to plants, an effect that is difficult to ascertain because the symptoms they case occur only over the long term, thus increasing even more the hazards associated with the use of contaminated materials in agricultural activities.

In a review about the environmental risks of adding phosphate fertilizers contaminated with trace elements in long-term, reference [87] warn about the possibility of increasing concentrations of As, Cd and Pb in agricultural soils using contaminated fertilizers. The authors also emphasize that to prevent the entry of potentially toxic elements in agriculture, it is necessary a strict control on the sources of macro and micronutrients used in agriculture. This demonstrates the importance and need for further studies related to the use of fertilizers contaminated with toxic metals in crops and the environmental impacts that these sources of nutrients can cause.

6. SUSTAINABLE STRATEGIES FOR SOIL REMEDIATION

The introduction of toxic chemicals by man in various environmental compartments (soil, water and atmosphere), the of controlled or uncontrolled elimination of wastes (accidental spills, mining processes and smelting of metallic ores, fertilizer application, sewage sludge and pig manure in agricultural soils) puts the ability to self-cleaning of contaminated ecosystems in doubt. Consequently, the pollutants that accumulate toxic metals are of concern in relation to human exposure of ecosystems and the potential impact [88].

Today there are several techniques for soil decontamination. It is worth mentioning that the choice of one or other technique, among other factors, also depends on the characteristics of the contaminant in question, between the different techniques is possible enumerate:

1 Thermal treatment: Use of heat to remove, stabilize or destroy the contaminants;

2 Physical treatment: Physical processes used to separate toxic substances from the environment;

3 Chemical treatment: Uses chemical reactions to remove, destroy or modify toxic substances;

4 Biological treatment: Makes use of natural metabolic means, such a microorganisms and other biological agents to remove, destroy or modify the contaminants;

5 Stabilization / solidification: This step the contaminants are stabilized chemically and / or modified to reduce the potential for contamination.

These technologies can in many cases be applied “in situ”, i.e., at the site of origin of the contamination, or “ex situ” outside the place of contamination, which results in the removal and transport of the contaminated soil.

Among the physical-chemical treatments, these mostly are based on soil washing. These methods are based on the technological principle of the transfer of a contaminant from the soil for a liquid or gaseous phase. The main products are treated soil and the contaminants concentrated. The specific process of treatment depends on the type of contaminant, in particular as regards the type of connection established with the soil particles.

It is observed by the description above, that such treatments become process more expensive the decontamination because are many cases in which the soil treatment requires the removal of several layers of the same and specific treatment in laboratories.

Thus, many approaches have been developed, evaluated and performed to deal with soil pollution. Current technologies for decontamination soil are often expensive and of high energy consumption, and in many cases, the soil cannot be reused after treatment [89]. Thus, several studies have been developed to promote proper techniques and low cost of remediation in order to prevent the spread of these contaminants in the food chain [88, 90].

As an alternative, an ecological approach technology that involves the use of plants to clean up or remediate soils contaminated with toxic metals, called phytoremediation, has been developed and encouraged by the fact of being a simple and inexpensive option for remediation of contaminated soils [90-92].

Phytoremediation can be divided into phytoextraction (hiperacumulation) phytostabilization, rhizofiltration, fitotransformation and fitovolatilization (Figure 4).

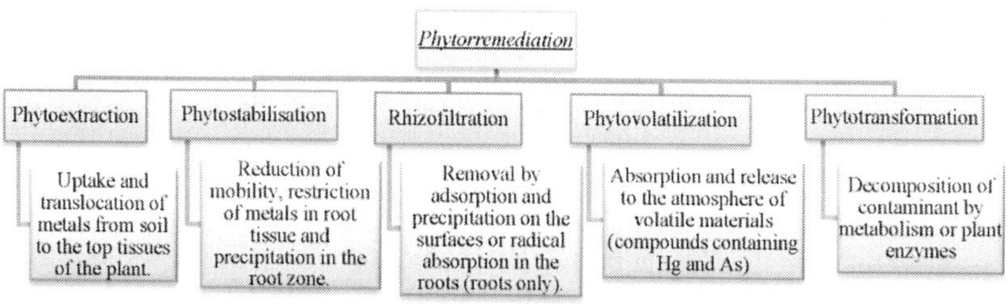

Figure 4. Types and characteristics of phytoremediation [92,93]

However for the phytoremediation to be effective it depends on the selection of plant species, preferably hyperaccumulators, which are able to grow in soils of low fertility, and degraded by the presence of contaminants [94].

In studies carried out in Iran using *Euphorbia macroclada* and *Centaurea virgata*, demonstrate that the sampled species not only were able to grow in soils highly contaminated by metals, as well as were capable of accumulating high concentrations of Zn, Mn, Cu, Pb and Fe, in this way these plants can be classified as hyperaccumulators, in other words, have adequate potential for phytoremediation of contaminated soils [95].

However, the metal accumulation by the plants is only efficient after the contaminant be removed from the soil, for example, by harvesting the plant matter. If most of the toxic metals are located on the shoots, the harvest can be performed using traditional farming methods. In general, it is necessary to harvest the plants before the fall of the leaves, or before his death and decomposition, so that contaminants do not disperse or return to the soil.

After harvesting, the biomass must be processed for extraction and recollection of the majority of metals. Alternatively, the volume and the weight of the biomass can be reduced by thermal processes, physical, chemical or microbial. In the case of burning plant, for example, the energy produced represents an economic value of the process and the ash may be treated as an ore, which may still be extracted the metallic contaminant (especially if the ashes are enriched in one or two metals).

Even Brazil showing great potential species for bioremediation and phytoremediation, the recovery of contaminated areas, due to the great biodiversity and climate, favoring biological processes in pollution treatment, knowledge about these species of plants and microbial communities in phytoremediation potential are still scarce [96].

In research carried out by reference [97], it was reported that some plants were able to remediate soils contaminated with metals such as *Brassica juncea* for phytoextraction of Cd and Cr^{6+}, Cu, Ni and Pb,*Avena sativa* for Zn, *Pisum sativum* L. and *Zea mays* L. for removal of Pb.

However, from an economic point of view, the use of these species of accumulator plants generally do not generate financial returns to the farmers, because these plants only remove the contaminant from the soil, being later discarded after the remediation. Thus if the farmer in question require to decontaminate a particular area, the phytoremediation only would generate costs, especially with seeds, agricultural machinery, irrigation, etc.

Nevertheless, currently there is a widespread idea that the species used for phytoremediation of soil can also generate economic gains beyond decontamination. The use of oil crops, for example, *Crambe abyssinica H., Moringa oleifera L.* and *Jatropha curcas*, in which the product marketing is not consumed as food, but as biofuel (biodiesel), is a possibility to decrease the cost of recovery of the contaminated area. Thus the fact that the plant has accumulated large amounts of heavy metal will have little influence on the end product of the culture, generating profit for the property.

Moreover, oil crops like *C. abyssinica, M. oleifera* and *J. curcas*, have been the subject of research in the removal of toxic metals such as Cd, Pb, Cr and water through the adsorption process, in which the pie their seeds are used as adsorbent [98-101]. The results of these studies demonstrated great potential for removal of these metals, and that these adsorbents are obtained at minimal charges, because they are wastes generated by the extraction of vegetable oil.

Thus, the use of cultures for vegetable oil production and recovery of contaminated areas can be an interesting alternative, which will generate the money from oil production and recovery / stabilization of the contaminated area with the development of these cultures, preventing them be lost through leaching or soil surface runoff and reach the groundwater and surface water and groundwater. However, more researches are still needed In order to evaluate this possibility, focusing on the development of oil-producing plants and cycles of addition of the adsorbent material and the biodegradability thereof. Between this and other issues to which scientific research is driven, many studies still need to be developed so that, in one way or another, the quest for sustainability will one day be achieved.

7. CONCLUSION

Based on information existing in the Brazilian and international literature, we can make some important observations:

The fertilization of plants is one of the major reasons of the increase in food production across the globe. Consequently, the use of fertilizers in agricultural crops is extremely high and growing, however, for the maintenance of this production capacity, we should be very careful with the fertilizers sources that are being used in agriculture.

The presence of toxic metals in fertilizers has been reported in the major food producing countries, like United States, China and Brazil. This demonstrates that this issue deserves further detailed studies of long-term about the successive addition and the effects of these contaminants to crops, soil, supply chain and the wider environment. Thus, the institutions of research and supervision, have to evaluate the products that are used for the manufacture of fertilizers on the market, as this is an activity that generate large amounts of money and that will always be a target of people with questionable character.

The current legislation, not only in Brazil, but internationally, with respect to the use of wastes as sources of nutrients, presence of contaminants in fertilizers and presence of toxic metals in soil, need higher accompaniments and answers for some of the community questions, because the legislation must ensure that the agricultural soils are not being degraded and becoming unproductive. In Brazil, it is strongly recommended that the laws must be harder and restrictive, until we have reliable and conclusive results related to the addition of contaminated fertilizer in long-term. In this case, it should be emphasized that these laws should be developed based on results from institutional researches and reliable groups, without the influence of political and industrial groups. Meanwhile, the damage caused by man in air, soil and waters, must be repaired, by remediation methods (such as phytoremediation) aiming not only the decontamination, but also ensuring a small profit, as an incentive to the farmer.

Brazil is considered as one of the countries with largest biodiversity and most promising in food production for the globe, however, for that prediction become a reality, it is very important that the productive areas are not degraded. The technologies developed by our civilization provide ways to avoid the degradation of our environment, but often they are not used because they prevent higher industrial profits, this should not occur, because the balance of the ecosystem and the quality of human life should be above of any monetary profit.

REFERENCES

1. Nacke H, Gonçalves Jr AC, Stangarlin JR,Schwantes D, Strey L, Nava I. Productivity and yield components of maize fertilized with different sources and doses of zinc. Spanish Journal of Rural Development 2011;2(1) 47-55.

2. Nava I, Gonçalves Jr AC, Nacke H, Guerini VL, Schwantes D. Availability of cadmium, lead and chromium toxic heavy metals in soil and soybean leaf tissue fertilized with different sources of NPK+Zn. Ciência e Agrotecnologia 2011;35(1), 884-892.

3. Gonçalves Jr AC, Nacke H, Marengoni, NG, Carvalho, EA, Coelho GF. Yield and production components of soybean fertilized with different doses of

phosphorus, potassium and zinc. Ciência e Agrotecnologia 2010;34(3) 660-666.

4. Gonçalves Jr AC, Nacke H, Schwantes D, Nava IA, Strey L. Phytoavailability of toxic heavy metals and productivity in wheat cultivated under residual effect of fertilization in soybean culture. Water, Air and Soil Pollution 2011;220(1) 205-211.

5. Nacke H, GonçalvesJr AC, Schwantes D, Nava I, Strey L, Coelho GF. Availability of heavy metals (Cd, Pb and Cr) in agriculture from commercial fertilizers. Archives of Environmental Contamination and Toxicology 2013;64(1), 371-379.

6. Gonçalves Jr AC, Luchese EB, Lenzi E. Evaluation of phytoavailability of the cadmium, lead and chromium in soybean cultivated in the Latossolo vermelho escuro, treated with commercial fertilizers. Química Nova 2000;23(2) 173-177.

7. Manzatto CV, Freitas Junior E, Peres JRR. Agricultural use of Brazilian soils. Rio de Janeiro: Embrapa Solos; 2002.

8. FAO - Food and Agriculture Organization of the United Nations. The state of food and agriculture: Women in agriculture closing the gender gap for development. Rome: FAO; 2011.

9. Nelson GC, Rosegrant MW, Palazzo A, Gray I, Ingersoll C, Robertson R, Tokgoz S, Zhu T, Sulser TB, Ringler C, Msangi S, You L. Food Security, Farming, and Climate Change to 2050: scenarios, results, policy options. Washington: International Food Policy Research Institute; 2010.

10. Foley JA, Ramankutty N, Brauman KA, Cassidy ES, Gerber JS, Johnston M, Mueller ND, O'Connell C, Ray DK, West PC, Balzer C, Bennett EM, Carpenter SR, Hill J, Monfreda C, Polasky S, Rockstrom J, Sheehan J, Siebert S, Tilman D, Zaks DPM. Londons: Solutions for a cultivated planet. Nature 2011;478(1) 337-342.

11. International Fertilizer Industry Association. http://www.fertilizer.org/ifa/ifadata/search (accessed 15 June 2013).

12. Yamada T. Deficiências de micronutrientes, ocorrência, detecção e correção: o sucesso da experiência brasileira. Piracicaba: Potafos; 2004.

13. Cakmark I. Plant nutrition research: Priorities to meet human needs for food in sustainable ways. Plant and Soil 2002;247(1) 03-24.

14. Gonçalves Jr AC, Pessoa ACS. Phytoavailability of the cadmium, lead and chromium in soybean cultivated in argissolo vermelho eutrófico, treated with commercial fertilizers. Scientia Agrária2002;3(1-2) 19-23.

15. CETESB - Environmental Sanitation Technology Company of São Paulo State. Relation of contamined areas in São Paulo 2012. http://www.cetesb.sp.gov.br/userfiles/file/areas-contaminadas/2012/texto-explicativo.pdf (accessed 15 June 2013).

16. Lopes JC, Ribeiro LG, Araújo MG, Beraldo MRBS. Lettuce production using doses of sewage sludge. HorticulturaBrasileira 2005;23(1), 143-147

17. Pegorini ES, Andreoli CV. Introduction. In: Andreoli, CV. (ed.) Alternative use of sanitary waste. São Paulo:ABES; 2006. p01-06.

18. Marques MO, Nogueira TAR, Fonseca IM, Marques TA. Cr, Ni, Pb e Zn contents in sewage sludge added Red Argisol and sugarcane cropped. Revista de Biologia e Ciência da Terra2007;27(1) 133-143.

19. Alamino RCJ, Polivanov H, Campos TMP, Silva VHG, Santos LV, Mendes JC. Bioavailability of Cadmium on a Latosol Increased of Sewage Sludge. Anuário do Instituto de Geociência 2007;30(2) 45-54.

20. CETESB - Environmental Sanitation Technology Company of São Paulo State. Report establishment of guiding values for soils and groundwater in the State of São Paulo. São Paulo: CETESB; 2001.

21. Mattias JL, Ceretta CA, Nesi CN, Girotto E, Trentin EE, Lourenzi CR, Vieira RCB. Copper, zinc and manganese in soils of two watersheds in Santa Catarina with intensive use of pig slurry. Revista Brasileira de Ciência do Solo 2010;34(4) 1445-1454.

22. Wu L, Tan C, Liu L, Zhu P, Peng C, Luo Y, Christie P. Cadmium bioavailability in surface soils receiving long-term applications of inorganic fertilizers and pig manure. Geoderma 2012;173-174(1) 224–230.

23. Ramalho JFGP, Sobrinho NMB, Velloso ACX. Heavy metals contamination of a watershed in Caetés by the use of agrochemicals. Pesquisa Agropecuária Brasileira 2000;35(7) 1289-1303.

24. Baker DE. Copper. In: Alloway, BJ. (ed.) Heavy metals in soils. Glasgow: Blackie and Son; 1990. p.51-74.

25. Nicholson FA, Smith SR, Alloway BJ, Carlton-Smith C, Chambers BJ. An inventory of heavy metals inputs to agricultural soils in England and Wales. The Science of the Total Environment 2003;311(1) 205–219.

26. Council Directive of 12 June 1986 on the protection of the environment, and in particular of the soil, when sewage sludge is used in agriculture (86/278/EEC). http://eur-lex.europa.eu/LexUriServ/LexUriServ. do?uri= CELEX:31986L0278:EN:HTML (accessed 15 July 2013).

27. Kabata-Pendias A, Pendias H. Trace elements in soils and plants. 3rd ed. Boca Raton: CRC Press; 2001.

28. Brazil. Resolution 420. Brasília: Ministry of the Environment; 2009.

29. Novais RF, Alvarez VVH, Barros NF, Fontes RLF, Cantarutti RB, Neves JCL. editors. Soil Fertilty: Viçosa: SBCS; 2007.

30. USEPA – United States Environmental Protection Agency. Background Report on Fertilizer Use, Contaminants and Regulations. Columbus: USEPA; 1999.

31. The Weinberg Group. Scientific Basis for Risk-Based Acceptable Concentrations of Metals in Fertilizers and Their Applicability as Standards. Washington: The Weinberg Group; 2001.

32. USEPA – United States Environmental Protection Agency. Zinc fertilizers made from recycled hazardous secondary materials. Washington: USEPA; 2002.

33. WDE – Washington Department of Ecology. Focus: Imported Cadmium-Contaminated Zinc Sulfate Used in Fertilizer and Other Products. Washington: WDA; 2000.

34. Gabe U, Rodella AA. Trace elements in Brazilian agricultural limestones and mineral fertilizers. Communications in Soil Science and Plant Analysis 1999;30(5-6) 605-620.

35. Martins ALC, bataglia OC, Camargo AO, Cantarella H. Corn yield and uptake of Cu, Fe, Mn and Zn from sewage sludge-amended soil with and without liming.Revista Brasileira de Ciência do Solo 2003;27(3) 563-574.

36. Brazil. NormativeInstruction 27. Brasília: Ministry of Agriculture, LivestockandSupply; 2006.

37. Hooda SP. Trace Elements in Soils. London: Wiley; 2010.

38. CCR – California Code of Regulations. Standards and labeling. California: CCR; 2007.

39. AAPFCO – Association of American Plant Food Control. AAPFCO's Statement of Uniform Interpretation and Policy (SUIP) #25: the heavy metal rule. http://www.aapfco.org/rules.html (accessed 15 June 2013).

40. AFPC – Association of Fertilizer and Phosphate Chemists. Latest Metal Regulations in Fertilizers. http://afpc.net/Metal%20Regulations.html (accessed 15 June 2013).

41. ABNT – Brazilian National Standards Organization. NBR 10004. Rio de Janeiro: ABNT; 2004.

42. Rodella AA. Regulation of contaminant contents in fertilizers: A Case study. Revista Brasileira de ciência do solo 2005; 29(1) 797-801.

43. Pino GAH. Biosorption of heavy metals using *Cocos nucifera*. Master's Dissertation. Pontifical Catholic University; 2005.

44. Morillo J, Usero J, Gracia I. Partitioning of metals in sediments from the Odiel river (Spain). Environment International 2002;28(4) 263-271.

45. Bueno BYM. Removal of Pb, Cr and Cu by a Combined Biosorption / Bioflotation process using a *Rhodococcusopacus* straint. Doctorate Thesis. Pontifical Catholic University; 2007.

46. Nriagu JO, Pacyna JM. Quantitative assessment of worldwide contamination of air, water and soils by trace metals. Nature 1988;333(1) 134–139.

47. Loska K, Wiechula D, Korus I. Metal contamination of farming soils affected by industry. Environmental International 2004;30(1) 159-165.

48. Naimo TJ. A review of the effects of heavy metals on freshwater mussels. Ecotoxicology 1995;4(6) 341-362.

49. Sayler GS, Nelson Jr JD, Colwell RR. Role of bacteria in bioaccumulation of mercury in the oyster *Crassostrea virginica*. Applied Microbiology 1975;30(1) 91–96.

50. Peralta-Videa JR, Lopez ML, Narayan M, Saupe G, Gardea-Torresdey J. The biochemistry of environmental heavy metal uptake by plants: Implications for the food chain. The international Journal of biochemistry & Cell Biology 2009;41(8-9) 1665-1677.

51. ATSDR – Agency for Toxic Substances and Disease Registry. Toxic Substances Index. http://www.atsdr.cdc.gov/az/a.html (accessed 15 June 2013).

52. Rahman MA, Hasegawa H, Lim PR. Bioaccumulation, biotransformation and trophic transfer of arsenic in the aquatic food chain. Environmental Research 2012;116(1) 118-135.

53. Cao X, MA LQ. Effects of compost and phosphate on plant arsenic accumulation from soils near pressure-treated wood. Environmental Pollution 2004;132(3) 435-442.

54. ATSDR – Agency for Toxic Substances and Disease Registry. The Priority List of Hazardous Substances. http://www.atsdr.cdc.gov/SPL/index.html (accessed 15 June 2013).

55. Brady JE, Humiston GE. General Chemistry. 2th ed. Rio de Janeiro: Livros Técnicos e Científicos; 1986.

56. ICdA. INTERNACIONAL Cadmium ASSOCIATION, Revision of the Battery Directive: Public Consultation, 2001.

57. PINO G.A.H., Biossorption of Heavy metals on green coconut powder. Master Dissertation, PUC-RJ, March, 113p. 2005.

58. Paoliello MMB, Chasin AAM. Ecotoxicology of lead and its compounds. Salvador: CRA; 2001.

59. ILZSG – International Lead and Zinc Study Group. Statistic of Lead and Zinc. 2010. http://www.ilzsg.org/static/home.aspx (accessed 15 June 2013).

60. Kreush MA. Assessment with proposals for improving the industrial process of recycling lead and indicating applicability to slag generated. Master's Dissertation. Federal University of Parana; 2005.

61. ATSDR – Agency for Toxic Substances and Disease Registry. Case studies in environmental medicine: lead toxicity. Atlanta: ATSDR; 1999.

62. Capitani EM. Diagnosis and treatment of lead poisoning in children and adults. Medicina 2009;42(3) 319-329.

63. Cervantes C, Campos-García J, Devars S, Gutiérrez-Corona F, Loza-Tavera H, Torres'Guzmán JC, Moreno-Sánchez R. Interactions of chromium with microorganisms and plants. FEMS Microbiology Reviews 2001;25(3) 335-347.

64. WHO - World Health organization. Environmental health criteria 61 - Chromium. Geneva: WHO; 1988.

65. Smith WLE, Gadd GM. Reduction and precipitation of chromate by mixed culture sulphate-reducing bacterial biofilms. Journal of Applied Microbiology 2000;88(6) 983-991.

66. Kumral E. Speciation of chromium inwaters via sol–gel preconcentration prior to atomic spectrometric determination. Doctorate Thesis. The Graduate School of Engineering and Sciences of Izmir Institute of Technology; 2007.

67. Frois SR, Grassi MT, Fernandes TC, Barreto RAS, Abate G. Preconcentration of Cr(III) and speciation analysis of chromium employing montmorillonite saturated with potassium ions. Química Nova 2011;34(3) 462-467.

68. Schaumlöffel D. Nickel species: analysis and toxic effects. Journal of Trace Elements in Medicine and Biology 2012;26(1) 1-6.

69. Veit MT, Silva EA, Fagundes-Klen MR, Tavares CRG, Gonçalves GC. Biosorption of nickel and chromium from a galvanization effluent using seaweed pre-treated on a fixed-bed column. Acta Scientiarum. Technology 2009; 31(2) 175-183.

70. Leonard A, Gerber P, Jacquet P. Carcinogenicity, mutagenicity and teratogenicity of nickel.Mutation. Research/Reviews in Genetic Toxicology 1981;87(1) 1-15.

71. Sodré FF, Lenzi E, Costa ACS. Applicability of adsorption models to the study of copper behaviour in clayey soils. Química Nova 2001;24(3) 324-330.

72. Lima AJB, Cardoso, MG, Guerreiro MC, Pimentel FA. Using activated carbon to remove copper from sugar cane spirit. Química Nova 2006;29(2) 247-250.

73. Yruela I. Copper in plants. Brazilian Journal of Plant Physiology 2205;17(1) 145-156.

74. International Manganese Institute. Applications of Mn. http://www.manganese.org/about_mn/applications (accessed 15 June 1013).

75. Das AP, Sukla LB, Pradhan N, Nayak S. Manganese biomining: A review. Bioresource Technology 2011;102(16) 7381-7387.

76. Zhang W, Cheng, CY. Manganese metallurgy review. Part I: Leaching of ores/secondary materials and recovery of electrolytic/chemical manganese dioxide. Hydrometallurgy 2007;89(3-4) 137-159.

77. Gerber GB, Leonard A, Hantson PH. Carcinogenicity, mutagenicity and teratogenicity of manganese compounds. Critical Reviews in Oncology/ Hematology 2002;42(1) 25-43.

78. Brault N, Loranger S, Courchesne F, Kennedy G, Zayed J. Bioaccumulation of manganese by plants: influence of MMT as a gasoline additive. Science of the Total Environment 1994;153(1-2) 77-84.

79. Google. Google Maps. https://maps.google.com (accessed 15 June 2013).

80. Paula Jr D. Office 2147/09-MP-PJCS-UMA. http://www.mma.gov.br/ port/ conama/processos/FE4582B1/Of2147-09-MP-PJCS-U MA_7maio09.pdf. 2009. (accessed 15 June 2013).

81. Varjabedian R, Leite CBB. Office 1029/2009 – PGJ/CAO. http:// www.mma.gov.br/port/conama/processos/FE4582B1/ConsideracoesMinis PublSaoPaulo.pdf. 2009 (accessed 15 june 2013).

82. Brazilian Ministry of the Environment. Process 02000.002955/2004-69. 2011. http://www.mma.gov.br/port/conama/processo.cfm? processo= 02000.002955/2004-69 (accessed 15 June 2013).

83. Gimeno-García E, Andreu V, Boluda R. Heavy metals incidence in the application of inorganic fertilizers and pesticides to rice farming soils. Environmental Pollution 1996;92(1) 19–25.

84. Ryser P, Sauder WR. Effects of heavy-metal-contaminated soil on growth, phenology and biomass turnover of *Hieraciumpiloselloides*. Environmental Pollution 2006;140(1) 52–61.

85. Jan FA, Ishaq M, Khan S, Ihsanullah I, Ahmad I, Shakirullah M. A comparative study of human health risks via consumption of food crops grown on wastewater irrigated soil (Peshawar) and relatively clean water irrigated soil. Journal of Hazardous Materials 2010;179(1-3) 612–621.

86. Jiao W, Chen W, Chang AC, Page AL. Environmental risks of trace elements associated with long-term phosphate fertilizers applications: A review. Environmental Pollution 2012;168(1) 44-53.

87. Ghosh M, Singh SP. A Review on Phytiremediation of heavy metals and utilization of its byproducts. Applied ecology and environmental research 2005;3(1) 1-18.

88. Lorestani B, Cheraghi M, Yousefi N. Phytoremediation Potencial of Native Plants Growing on a Heavy Metals Contaminated soil of Copper mine in Iran. World Academy of Science, Engineering and Technology 2011;53(1) 377-382.

89. Susarla S, Medina VF, McCutcheon SC. Phytoremediation: An ecological solution to organic chemical contamination Ecological Engineering 2002;18(1) 647–658.

90. Gratão PL, Prasad MNV, Cardoso PF, Lea PJ, Azevedo RA. Phytoremediation: Green technology for the clean up of toxic metals in the environment. Brazilian Journal of Plant Physiology 2005;17(1) 53-64.

91. Lorestani B, Cheraghi M, Yousefi N. Accumulation of Pb, Fe, Mn, Cu and Zn in plantsand choice of hyperaccumulator plant in the industrial town of viean, Iran. Archives of Biological Science Belgrade 2011; 63(3) 739-745.

92. Jing Y, He Z, Yang X. Role of soil rhizobacteria in phytoremediation of heavy metal contaminated soils Journal of Zhejiang University Science B 2007;8(3) 192-207.

93. Mkumbo S, Mwegoha W, Renman G. Assessment of the phytoremediation potential for Pb, Zn and Cu of indigenous plants growing in a gold mining area in Tanzania. International Journal of Environmental Sciences 2012;2(4) 2425-2434.

94. Cheraghi, M,Lorestani, B,Yousefi, N. Introduction of Hyperaccumulator Plants with Phytoremediation Potential of a Lead- Zinc Mine in Iran. World Academy of Science, Engineering and Technology 2011, 53(1) 163-168.

95. Marques M, Aguiar CRC, Silva JJLS. Technical challenges and social, economic and regulatory barriers to phytoremediation of contaminated soils. Revista Brasileira de Ciência do Solo 2011;35(1) 1-11.

96. Prasad MNV, Freitas HMO. Metal hyperaccumulation in plants – Biodiversity prospecting for phytoremediation technology.Eletronic Journal of Biotechnology 2003;6(3) 285-321.

97. Gonçalves Jr. AC, Rubio F, Meneghel AP, Coelho GF, Dragunski DC, Strey L. The use of *Crambe abyssinica* seeds as adsorbent in the removal of metals

from waters. Revista Brasileira de Engenharia Agrícola e Ambiental 2013;17(2) 306-311.

98. Rubio F, Gonçalves Jr. AC, Meneghel AP, Tarley CRT, Schwantes D, Coelho GF. Removal of cadmium from water using by-product *Crambe abyssinca* Hochst seeds as biosorbent material. Water Science and Technology 2013;68(1) 227-233.

99. Rubio F, Gonçalves Jr. AC, Strey L, Meneghel AP, Coelho, GF, Nacke, H. Applicability of *Crambe abyssinica* Hochst byproduct as biosorbent in the removal of chromium from water. Spanish Journal of Rural Development. 2013;4(1) 25-40.

100. Gonçalves Jr. AC, Meneghel AP, Rubio F, Strey L, Dragunski DC, Coelho GF. Applicability of *Moringa oleifera* Lam. pie as an adsorbent for removal of heavy metals from waters. Revista Brasileira de Engenharia Agrícola e Ambiental 2013;17(1): 94-99.

101. Nacke H, Gonçalves Jr. AC, Coelho GF, Strey L, Schwantes D, Laufer A. Removal of cadmium from aqueous solutions by adsorption on *Jatropha biomass* In: Bartolo H. (ed.) Green Design, Materials and Manufacturing Processes. Boca Raton: CRC Press Taylor & Francis Group; 2013. p367-372.

Chapter 9

Reflectance Spectroscopy as a Tool for Monitoring Contaminated Soils

Guy Schwartz[1], Gil Eshel[3] and Eyal Ben-Dor[2]

[1] Tel Aviv University, Porter School of environmental studies, Israel

[2] Tel Aviv University, Geography and human environment department, Israel

[3] Soil Erosion Research Station, Ruppin Inst., Emeck-Hefer, Israel

1. INTRODUCTION

Soil reflectance spectroscopy is a well-known technique to assess soil properties rapidly and quantitatively in both point (spectroscopy) and spatial (imaging spectroscopy (IS)) domains. The quantitative approach has been developed in the past two decades by many researchers, with much literature on the topic. Basically, this approach was adopted from other disciplines (e.g. food, textiles), whereas the mapping approach was incorporated with spectral-based remote sensing means. Soil contamination is an ever-growing concern, and a reliable and environmentally friendly method is needed to rapidly detect and monitor soil contaminants at any stage for the diagnosis of suspected contaminated areas, as well as for controlling rehabilitation processes. Conventional methods for investigating soil contamination based on point sampling and chemical analysis are time-consuming, costly and sometimes incorporate the use of environmentally damaging chemicals. Soil reflectance spectroscopy is a promising tool for the efficient detection and monitoring of soil contaminants. Using reflectance spectroscopy measurements, several direct and indirect soil properties, as well as soil contamination characteristics, can be extracted. Due to the vast amount of data in soil spectroscopy, be it point or image, data-modeling processes are required. This chapter reviews soil spectroscopy theory and its quantitative capabilities, as well as data-modeling methods for soil spectra that are related to soil contamination. Discussed contaminants include petroleum hydrocarbons, heavy metals such as Ni, Cr, Cu, Cd, Hg, Pb, Zn and As, acid mine drainage and pesticides. Limitations, obstacles and problems in recording soil spectra in the field and in the laboratory are discussed, along with IS technology. The future potential of spectral technology and IS missions used for soil contamination monitoring is presented, based on ongoing worldwide initiatives. This chapter also provides some notes on how to merge this technology with other systems, such as a penetrating spectral device for underground measurements utilizing the catheterization approach.

1.1. ENERGY MATTER INTERACTIONS

Most, if not all matter interacts with electromagnetic radiation in one way or another. Electromagnetic radiation that is radiated onto any kind of material may be reflected, absorbed or transmitted. For most materials, specific light interactions of a certain wavelength can be observed in the same range as solar radiation (Hunt 1980). For that reason, this chapter focuses on this radiation range (VNIR-SWIR, 400 to 2,500 nm). This range is divided into two sub-ranges: VNIR (400 to 1,100 nm) and SWIR (1,100 to 2,500 nm). Interactions between light and matter are wavelength-dependent. While electromagnetic radiation radiated onto a specific material may be reflected at one wavelength, it might be absorbed at another. This means that the reflected radiation of a material is a function of wavelength, surface material absorption, emissivity and reflectance characteristics. Therefore, energy-matter interactions in the VNIR, SWIR and IR (2,500 to 12,000 nm) wavelength regions can be used for qualitative and quantitative chemical analysis of a material.

1.2. Spectroscopy

VNIR-SWIR and IR spectroscopy has been used for qualitative and quantitative analysis for several decades (Taylor et al. 1970; Stark et al. 1986). Both are based on characteristic vibrations of chemical bonds in molecules, as well as electronic processes (the latter affects only radiation at 400 to 1,000 nm) (Ben-Dor et al. 1999). In general, incident IR is absorbed by the chemical-bond vibrations in the molecules (3N-6 fundamental possibilities where N is the number of atoms in a given molecule). The combination modes and overtones of the fundamental vibrations in the IR region are observed in the VNIR-SWIR region, and can be detected if the incident radiation in this latter region interacts with the matter in question. Whereas using the IR region in the laboratory requires sample preparation and caution during the measurements (e.g. not to exceed the saturation point), the VNIR-SWIR region has the advantage of allowing direct measurement of the material without complicated pretreatment and reagents. Another advantage of VNIR-SWIR over IR spectroscopy is that it is a non-destructive and non-contact method: it allows real-time measurements based on a simple illumination source (e.g. sun, tungsten halogen bulb). It is therefore suitable for on-line, *in-situ* monitoring and analysis of many types of compounds, mixtures and materials. In contrast, IR spectroscopy is a laboratory analysis method that requires considerable sample preparation. Its advantage over VNIR-SWIR lies in the fact that all chemical bonds within the molecule are active in this region, whereas in the VNIR-SWIR region only some of the bonds are active (e.g. OH-, NH-). Another benefit of IR is that it can be interpreted directly based on known chemical bonds, while in the VNIR-SWIR region, the spectra must be studied in advance to evaluate which wavelengths will best represent a given property (this is due to the possible overlap between some combination modes and overtones in the studied material). As a result, complex chemometric processes are used for the interpretation of VNIR-SWIR spectra into a quantitative model, in a process termed NIRA or NIRS (near infrared analysis). In this chapter, we will focus on hyperspectral data (continuous spectral data, amounting to hundreds of spectral bands), in the field of soil-contamination monitoring by reflectance spectroscopy and NIRS.

1.3. Soil Spectroscopy

Today, the NIRS technique is well-known and utilized in many disciplines, such as food science, pharmacology, textile, tobacco, and oil industries, among others. Since the mid-1960s, when Bower and Hanks showed that soil moisture is highly correlated with soil spectra (Bowers & Hanks 1965), soil has captured the attention of many researchers who have realized that despite its complexity, soil spectroscopy can provide remarkable amounts of quantitative information. Stoner et al. (1980) were the first scientists to systematically gather soil spectral information and publish it in the form of a soil spectral atlas. Their soil spectral library very soon became a classic tool that soil scientists came to rely on. Later, when laboratory and portable field spectrometers were introduced into the market (around 1993), more scientists realized the potential of soil spectroscopy, and consequently more spectral libraries were assembled and new quantitative, chemometric applications, such as NIRA, were developed and implemented for various soil materials. A summary of soil reflectance theory and its applications can be found in various publications (Irons et al. 1989; Ben-Dor et al. 1999; Ben-Dor 2002). A study by Brown et al. (2006) showed that NIRA can work successfully under a generic global view rather than relying on a specific soil population. An extensive study with over 1,100 soil samples, using several data-mining algorithms to model and interpret diffuse reflectance spectra of soils and predict a number of soil attributes was recently performed by Rossel & Behrens (2010).

1.4. Imaging Spectroscopy and Remote Sensing

Imaging spectroscopy (IS), or hyperspectral remote sensing, is an advanced tool that generates data of high spectral resolution, with the aim of providing near-laboratory-quality reflectance (or emittance) for each single picture element (pixel) from a far distance (Vane et al. 1984). This information enables the identification of objects based on the spectral absorption features of chromophores and has been found very useful in many terrestrial and marine applications (Clark & Roush 1984; Dekker et al. 2001). IS brings a new dimension to the field of remote sensing by expanding the envelope of point spectrometry to the spatial domain. It provides a tangible perspective by adding spatial detail to spectral information, thereby enhancing the thematic application of spectral recognition algorithms. This capability exists for both far and close distances, such as data acquired by satellites or by microscopic sensors, respectively. Whereas the former is used for mapping the earth from space, the latter is used for mapping micro targets, such as microorganisms and cell bodies, to account for their biochemical processes in a spatial domain (Soenksen et al. 1996; Levenson & Farkas 1997). It is interesting to note, however, that although soil scientists have recognized the potential of reflectance spectroscopy and in fact termed it a novel technology, in many ways, the use of IS for soil applications remains undeveloped and is seldom reported. Though the IS approach is a cost-effective method, its adoption is limited because the data is difficult to process, only a few sensors are operated worldwide, and it has not yet been recognized by many end users. Hence, the journey from point spectroscopy to a cognitive (imaging) spectral view of soils has not yet been completed, although there is no doubt that it may open up new frontiers in the field of soil science. Thus far, only exclusive and select groups around the world have been able to use IS for soil applications. Nevertheless, over the past 10 years, these groups have demonstrated remarkable achievements and have documented its significant capability. For further information refer to the chapter on "Optical remote sensing techniques for soil contamination monitoring and vulnerability assessment" in this book.

1.5. Acquiring Reflectance Information From Soil

To acquire reflectance information from the VNIR-SWIR region, several sensors, methods, protocols and platforms are used. Basically, the VNIR-SWIR sensors consist of an apparatus (with filters, grating, prisms, interferometer, etc.) that splits the measured radiation into individual wavelengths (or regions), several detectors (mostly CCDs) which are sensitive to different regions), and various other optional optical components (lenses, collimators, beam-splitters, integrating spheres, optical fibers, etc.). The radiation source is usually a tungsten halogen light bulb when recording spectra under laboratory conditions, and the sun when recording spectra in the field. These sensors are generally periodically calibrated to provide the radiance values but in practice, the reflectance is calculated against a white reference target (Spectralon) that is assumed to reflect 100% of the radiation in the sample geometric configuration. There are point and image sensors. The point sensor records a single spectrum of the target (with a line-array detector) while averaging the mixed information of all components within the sensor's field of view. The size of the single pixel being measured depends upon the optics and the distance from the target. The image sensor provides data similar to the point sensor except that an area-array detector is used that enables the acquisition of spectral-cube information constructed from spatial and spectral views of the area (target) in question. Image sensors usually have lower spectral resolution as well as lower signal-to-noise ratio, as the integration time over the targets is less than in the point spectrometer measurements. Laboratory sensors are more accurate, with higher spectral resolution and a better signal-to-noise ratio: the integration time is long and the geometric scheme is constant. They usually include a sample holder and a stable light source. Field sensors are generally battery-operated and are less accurate than laboratory sensors. They rely on the sun's radiation and the geometric measurement is not fixed. Nonetheless, they provide flexibility and *in-situ* measurement capabilities that cannot be achieved under laboratory conditions. The field sensors are a valuable tool for recording reference spectra for space- or airborne remote sensing imagery which will be later used for calibration, validation and quality control (Brook & Ben-Dor 2011). Image sensors can be operated from both ground and air (space) domains. They can be integrated with the aircraft's GPS/INS system and produce multi or hyperspectral data cubes that can be georeferenced; thus each pixel in the image corresponds to a single area unit on the ground and is represented by a spectrum. The signal-to-noise ratios of air- and spaceborne sensors are lower than those operated on the ground and are governed by many factors (e.g. pixel size, integration time, frame rate, electronic noise, IFOV and atmosphere attenuation). A comprehensive review of image-based sensors for soil application can be found in (Ben-Dor et al. 2008).

2. QUANTITATIVE APPLICATIONS OF SPECTROSCOPIC DATA

Spectroscopic data are multivariate in nature. There are two approaches: 1) the chromophore absorption band in question is known and under saturation (supervised), and 2) the chromophore absorption band is unknown and is affected by several factors (unsupervised). Whereas the first approach is valid for the IR region by using selected wavebands and pretreatment to prevent saturation, the second represents the case of reflectance spectroscopy across the VNIR-SWIR region. For that purpose, multivariate statistical techniques (also called chemometrics) are required to extract the information about the quality attributes that is hidden within the spectral information. Essentially, this involves regression techniques coupled with spectral preprocessing.

2.1. Preprocessing

Spectral preprocessing techniques are used to remove any irrelevant information which cannot be handled properly by the modeling techniques. The preprocessing techniques include averaging, centering, smoothing, standardization, normalization and transformations, among others. Some of the most common preprocessing techniques are presented here. For a full review of preprocessing methods, please refer to Nicolaï et al. (2007).

2.1.1. Multiplicative Scatter Correction

Multiplicative scatter correction (MSC) is one of the most commonly used normalization techniques. In MSC, the light scattering is estimated for each sample relative to an ideal sample obtained by averaging the complete wavelength range of the data set. Each spectrum is then corrected such that all samples appear to have the same scatter level as the reference spectrum. For a further description of this method, see (Isaksson & Naes 1988; Naes et al. 1990; Vohland et al. 2009).

2.1.2. Absorbance

Spectroscopic measurements performed in transmission mode can be quantified using Beer's Law (concentration ~ $\log(1/T)$ where T is the transmitted radiation). Accordingly, reflectance measurements are frequently converted to log (1/*R*) values, which are then used in a manner similar to optical density readings (Nicolaï et al. 2007).

2.1.3. Kubelka-Munck

The Kubelka-Munck transformation, $(1-R^2)/2R$, is similar to absorbance but accounts for scattering (Nicolaï et al. 2007).

2.1.4. Continuum Removal

The continuum is the background absorption on which other absorption features are superimposed. The spectrum is divided by a convex hull that is fitted over it (Clark & Roush 1984; Clark & others 1999).

2.1.5. Derivatives

Derivation is often used to remove baseline shifts and superposed peaks. Second-derivative spectra can correct for both additive and multiplicative effects (like MSC). They are usually calculated according to the Savitzky-Golay algorithm (Naes et al. 2002). The parameters of the algorithm (interval width, polynomial order) should be carefully selected to avoid amplification of spectral noise (Nicolaï et al. 2007).

2.1.6. Spectral Mixture Analysis

Spectral mixture analysis (SMA) is a widely used method to determine the sub-pixel materials that fundamentally contribute to the spectral signal of mixed pixels. This is of particular importance for obtaining quantitative estimates of distinct materials, a typical application of remote sensing hyperspectral data. SMA aims to decompose the measured reflectance spectrum of each pixel into the proportional spectral contribution of so-called endmembers (EMs). In recent years, many authors have proposed and used a more complex model, in which both the number and the set of EMs vary dynamically on a per-pixel basis; this has become known as multiple EM SMA (MESMA). The idea consists of restricting the large set of possible EMs to a small set of better suited EMs, which can be different for each pixel, thereby allowing an accurate decomposition using a virtually unlimited number of EMs. An improved strategy, called variable MESMA (VMESMA), allows segmentation of the image to increase flexibility and accuracy (Kemper & Sommer 2003).

2.2. Modeling

Modeling refers to relating a set of spectral parameters that are derived from the spectral information (before or after the aforementioned preprocessing treatment), to the real chemistry of the material in question. This is done by using a set of well-known samples as a training group. The data are divided into three groups: training, validation and test. The relationship between the chemistry and the spectroscopy data is found via the training group and simultaneously cross-validated by the validation group. Finally, the model is applied to the test group, independent of the training and validation process. Multivariate regression techniques are modeling methods that search for the relationship between two matrices: the spectral data matrix that can be very complex due to large amounts of data (X variables, the independent data), and a specific chemical reference value data matrix (Y variables, the dependent data). The common multivariate regression techniques are presented herein. For an in depth reference, please read further at (Esbensen et al. 2002; Nicolaï et al. 2007).

2.2.1. Multiple Linear Regression

Multiple linear regression (MLR) is a classical method that creates a linear combination of the spectral values at every single wavelength to correlate as closely as possible to the dependent reference values. The regression coefficients are estimated by minimizing the error between predicted and observed response values in a least squares sense. MLR models typically do not perform well with spectral data because spectral data usually exhibit high co-linearity, noise and more variables, i.e. more spectral bands, than measured samples (Esbensen et al. 2002; Nicolaï et al. 2007).

2.2.2. Principal Component Regression

Principal component regression (PCR) is a combination of principle component analysis (PCA) and MLR. The independent data matrix (spectral data) is transformed by PCA, and the first few principal components (PCs), which represent most of the independent data variance, are used as inputs for the MLR model instead of the original spectral data. The advantage over standard MLR is that PCs are uncorrelated, and the noise is filtered. The first few PCs are usually sufficient for a robust model and over-fitting issues can be eliminated. Although PCR consists of the two most studied multivariate methods (PCA and MLR), the major criticism against it is that the PCs (several first PCs) selected for the MLR input are not necessarily the best predictors for the reference data. There is no guarantee that the first PCs will include the spectral data related to the specific dependent variable that needs to be modeled (Esbensen et al. 2002; Nicolaï et al. 2007).

2.2.3. Partial Least Squares Regression

Introduced in 1983 by Wold et al., partial least squares regression (PLS) is similar to PCR, but in PLS the PCs are constructed such that they include the chemical reference (Y variables, dependent data) in the calculation process. This technique orders the PCs according to their relevance for predicting the dependent variables, rather than to their description of the most variance of the spectral data. This method excels when the dependent data (X variables) express common information, as usually happens in spectral data. The required number of PCs is typically smaller than that in a PCR calibration model for similar model performance (Wold et al. 1983; Esbensen et al. 2002; Nicolaï et al. 2007).

2.2.4. Artificial Neural Networks

Artificial neural networks (ANNs) are based on their ability to "learn" during a training process where they are presented with inputs and a set of expected outputs. The basic structure of an ANN consists of three layers of "nodes" or "neurons": an input layer (i.e. spectral data or PCs), a hidden layer (which can consist of one of more nodes), and an output layer (which combines the outputs of the hidden layer into a single output). The node is a fundamental processing unit; each node has a series of weighted inputs, either from an external source or the output from other nodes. The inputs to the node are analogous to synapses, and the weights correspond to the strength of the synaptic connection. The sum of the weighted inputs is transformed with linear or nonlinear transfer functions, a popular nonlinear transformation function being the sigmoid function. The learning (or training) is an iterative process in which the resultant output is compared to the provided expected output, and an algorithm adjusts the weights accordingly. This method was first tested in the field of spectroscopy on simulated data (Long et al. 1990). It was proven to provide the best results in modeling soil constituents in a recent large-scale study that included over 1,100 soil samples (Rossel & Behrens 2010).

3. HEAVY METALS

3.1. Introduction

Heavy metals are released into the environment—the air (e.g. during combustion, extraction and processing), surface water (via runoff and release from storage and transport) and soil (and hence into the groundwater and crops). Although the adverse health effects of heavy metals have long been known, exposure to heavy metals continues and is even increasing in some areas, especially in less developed countries. The main threats to human health from heavy metals are associated with exposure to Pb, Cd, Hg and As. Heavy metal exposure may cause headaches, irritability, abdominal pain, kidney damage, skeletal damage, acute pulmonary effects, cardiovascular disease, chronic renal failure, cancer, lung damage, neurological and psychological symptoms, nervous system damage and much more (Järup 2003). Heavy metals as soil contaminants typically occur in low concentrations (10,000 mg kg^{-1} dry soil). Inorganic material in general does not exhibit characteristic absorption features in the VNIR-SWIR wavelength region. It is therefore considered impossible to directly detect inorganic soil contaminants in general, and heavy metals in particular (Winkelmann 2005). The conventional method of estimating the spatial distribution of heavy metals is by raster sampling and a time-consuming laboratory analysis, followed by geostatistical interpolation (Kemper & Sommer 2002). While pure metals do not absorb in the VNIR-SWIR region, they may be detected indirectly via their complexing with organic matter (OM), association with moieties such as hydroxides, sulfides, carbonates, or oxides which are detectable, or adsorption to clays that absorb light in this wavelength range (Malley & Williams 1997). To explain possible variations in spectral signals due to heavy metals bound to minerals, it is necessary to consider the binding reaction of the metal onto the mineral surface. This approach starts from the premise that the spectrally assigned position of minerals can change with chemical composition and surface activity (Ben-Dor et al. 1999). Despite the occurrence of otherwise similar minerals in different samples, variations in spectral features (e.g., shifts in peak wavelength) may occur according to the nature of highly enriched cations in the mineral. The surface complexation model of the binding reactions of heavy metals describes the binding of metal ions to the mineral surface functional group to form a more stable molecular unit (Christl & Kretzschmar

1999;Zachara & Westall 1999). The main surface functional groups are inorganic hydroxyl groups that bind to surface Al, Fe, Mn, or Si on oxides or Al and Si on the edges of clay minerals (Sparks 1995;Zachara & Westall 1999; Choe et al. 2008).

3.2. Reflectance Spectroscopy Of Heavy Metals

The first report of quantitative prediction of heavy metals in sediments by NIRS was published in 1997 by Malley and Williams. They showed the feasibility of predicting six of the seven metals examined (Cd, Cu, Pb, Zn, Ni, Mn, Fe) with NIRS in a highly variable set of sediment samples, collected at Precambrian Shield Lake in northwestern Ontario, Canada. The metals were modeled by both MLR and PLS, and OM was found to be responsible for the prediction of sediment heavy metal concentrations (Malley & Williams 1997). A study published in 2002 by Kemper and Sommer investigated the ability to predict heavy metals following a mining accident that caused an area to be flooded with pyritic sludge contaminated with high concentrations of heavy metals: 214 soil samples were collected, and prediction of heavy metals was achieved by MLR and ANN approaches. It was possible to predict six out of nine elements (As, Fe, Hg, Pb, S, Sb) with high accuracy. Correlation analysis revealed that most of the wavelengths that were important for the prediction could be attributed to absorption features of Fe and Fe oxides (Kemper & Sommer 2002). The study by Kemper and Sommer continued with the collection of an additional 132 soil samples and acquisition of hyperspectral data with the HyMap airborne sensor (Cocks et al. 1998). A spectral mixture modeling approach was applied to the field and airborne hyperspectral data. VMESMA was used to estimate the quantities and distribution of the remaining tailing material. The semi-quantitatively determined abundance of residual pyritic material in the sludge could be transformed into quantitative information for an assessment of acidification risk and distribution of residual heavy metal contamination based on an artificial mixture experiment prepared with three different soils and pure sludge. Unmixing of the HyMap images allowed identification of secondary minerals of pyrite as indicators of pyrite oxidation and associated acidification (Kemper & Sommer 2003). The first study on NIRS of heavy metal contamination in agricultural soils was reported by Wu et al. (2005): 120 soil samples were collected in the Nanjing region, and Hg concentration was modeled by PCR. Correlation analysis revealed that Hg concentration is negatively correlated with soil reflectance, while positively correlated with the absorption depths of goethite at 496 nm and clay minerals at 2,210 nm, suggesting Hg sorption by clay-size mineral assemblages as the mechanism by which to predict spectrally featureless Hg in soils (Wu et al. 2005). In later work, Wu et al. (2007) also explored a physicochemical mechanism that allows estimation of heavy metals with the reflectance spectroscopy method, and concluded that correlation with total Fe (including active and residual Fe) is the major mechanism (Wu et al. 2007).Choe et al. (2008) explored the possibility of extending the use of spectral variations associated with heavy metal parameters to map the distribution of areas affected by heavy metals on HyMAP data in the Rodalquilar gold-mining area in southeast Spain. The chosen spectral parameters showed significant correlations with concentrations of Pb, Zn and As. Later, Choe et al. (2009) examined the potential use of spectral absorption feature parameters to quantify heavy metals using stepwise MLR (SMLR) and enter MLR (EMLR).

Contradictory conclusions on the major absorption feature mechanisms that can enable indirect prediction of heavy metals were presented by the abovementioned studies. However, their results cannot be generalized, as soil properties (e.g., soil C_{org} content) were distinctly different in the different studies (Vohland et al. 2009). Vohland et al. (2009) used soil samples taken from grassland plots that showed a very wide range of both C_{org} (0.37%–9.24%) and clay contents (0%–37.8%), with the aim of investigating whether the heavy metal contents of these samples can actually be assessed by NIRS. In addition, they explored statistically whether a valid prediction mechanism can be identified. According to the correlation spectra and PLS coefficients, it was concluded that the active C_{org}fraction is the key variable for an indirect spectroscopic approach (Vohland et al. 2009). Another study investigating the binding form of heavy metals with agricultural soil constituents was conducted by JiA et al. (2010). The results showed that Cr, Cu, Zn and As have stronger negative correlation coefficients with the absorption features of Fe oxides, clays and OM, However, Cd, Pb and Hg only displayed a significant correlation with the spectral region related to organic carbon (JiA et al. 2010). Pandit et al. (2010) showed very high correlation coefficients between laboratory-determined abundance and the abundance predicted from PLS calibration for several heavy metals, with the correlation coefficient for Pb being the highest of all metals (R=0.992). Soil Pb contamination exhibited a weak correlation (0.473) with percentage of OM, suggesting that while OM is a factor in Pb retention, it cannot be used to determine the quantity of Pb because only a little OM is needed to retain Pb (Pandit et al. 2010).

3.3. Conclusion

Heavy metal elements cannot be detected with reflectance spectroscopy at concentrations under 1000 mg kg^{-1} (Wu et al. 2007). The presence of Fe in ferric or ferrous forms results in absorption features at wavelengths in the 0.4 to 1.3 μm region. A doublet absorption feature near 496 nm is caused by goethite. Two distinct absorption bands at 1,410 and 1,910 nm are attributable to vibrational frequencies of OH groups in the adsorbed water, and the absorption feature at around 2,210 nm is related to OH groups in the crystal lattice water. For pure goethite, two absorption bands, at 670 and 950 nm, are also characteristic and even stronger than the 496 nm band. However, the soil matrix effect prevents these two bands from being displayed in soil spectra (Wu et al. 2005). The intercorrelation between heavy metals and total Fe (including active and residual Fe) is a major predictive mechanism (Wu et al. 2005). Later, Wu et al. (2007) showed that all heavy metals that have a high correlation with Fe also have high cross-validation statistics. The three siderophile elements Ni, Cr, and Co had the highest correlation coefficients with Fe, and prediction accuracy for these elements was also the highest among the eight elements studied. The most poorly predicted element, Cd, showed the lowest correlation with Fe. Another important soil constituent, OM, could absorb electromagnetic waves and adsorb heavy metals; however, the correlation between OM and heavy metals was very poor in the study area (Wu et al. 2007). This is different from results obtained using reflectance spectroscopy in food and forage crop research, where the content of inorganic elements is low relative to the organic matrix. Thus, successful prediction of inorganic components in forage crops can be based on the correlations between inorganic components and organic constituents (Siebielec et al., 2004). Choe et al. (2008) showed that despite the requirement for further corrections, the considered spectral absorption feature parameters had the potential to detect heavy metals. Moreover, the image-derived spectral parameters themselves showed a capacity to screen areas affected by heavy metals as a preliminary observation in determining sampling strategies and precise analyses for the investigation of environmental pollution (Choe et al. 2008). While weak prediction performance might result from the occurrence of many overlapping bands and the limited number of molecules that respond spectrally in the VNIR-SWIR range, a spatial distribution map of heavy metals by EMLR shows a distribution pattern similar to that on a map of measured values (Choe et al. 2009). However, the partial trade-off for time saved is reduced accuracy. Thus, spectral assessment of soil samples cannot replace, but rather complements the classical chemical analysis. The benefits are the practicable processing of a large number of samples and the savings on chemicals, lengthy and tedious processes and manpower. However, a calibration using information obtained from chemical analyses is mandatory (Vohland et al. 2009). In terms of spatial analysis, an airborne or space borne hyperspectral sensor may be useful for the screening of large areas and reproduction of the spatial distribution patterns.

3.4. Summary Table

The results of the previously mentioned NIRS prediction models of heavy metals are summarized inTable 1. Different modeling techniques as well as various preprocessing methods were employed in the development of these models, for an in depth description of the models please refers to the proper references.

Table 1. Heavy metals prediction via NIRS.

Authors	Contaminant (R^2)	# of Samples	Modeling Method
Malley & Williams (1997)	Cd(0.63), Cu(0.91), Fe(0.86), Mn(0.93), Ni(0.88), Pb(0.81), Zn(0.93)	169	MLR / PLS
Kemper & Sommer (2002)	As(0.84), Cd(0.51), Cu(0.43), Fe(0.72), Hg(0.96), Pb(0.95), S(0.87), Sb(0.93), Zn(0.24)	214	MLR / ANN
Kemper & Sommer (2003)	As(0.73), Pb(0.73)	346	polynomial fit
Siebielec et al. (2004)	Cd(0.54), Cu(0.61), Fe(0.87), Ni(0.84), Pb(0.45), Zn(0.67)	70	PLS
Wu et al. (2005)	Hg(0.48)	120	PCR
Wu et al. (2007)	As(0.72), Cd(0.20), Co(0.80), Cr(0.85), Cu(0.67), Ni(0.81), Pb(0.55), Zn(0.56)	61	PLS
Choe et al. (2008)	As(0.88), Pb(0.61), Zn(0.60)	49	EMLR
Choe et al. (2009)	As(0.60), Cu(0.81)	22	EMLR
Ren et al. (2009)	As(0.62), Cu(0.41), Fe(0.78)	33	PLS
Vohland et al. (2009)	Cu(0.75), Fe(0.84), Mn(0.71), Pb(0.76), Zn(0.81)	149	PLS
Jia et al. (2010)	As(0.30), Cd(0.10), Cr(0.68), Cu(0.46), Hg(0.15), Pb(0.26), Zn(0.40)	122	Linear Regression
Pandit et al. (2010)	Cd(0.43), Cu(0.81), Mn(0.81), Pb(0.75), Zn(0.76)	8	PLS

4. PETROLEUM HYDROCARBONS

4.1. Introduction

Petroleum hydrocarbons (PHCs) are environmental contaminants of great significance. With production rates of 13.42 million cubic meters of crude oil per day (Energy Information Administration 2009), their potential as soil and water contaminants is apparent. Crude oil and refined crude oil products therefore appear in almost every location on earth at the different stages of crude oil production, handling, transport, refining and consumption. The widespread use of PHCs for transportation, heating and industry has led to the release of these fuels into the environment through accidental spills, long-term leakage, and operational failures. Consequently, many soil and water areas are contaminated with PHCs.

PHCs are well known to be neurotoxic to humans and animals. They have been found to affect brain activity and development, and cause many other symptoms including, but not limited to: nausea, disorientation, mental confusion, speech slurring and memory disorders. Exposure to higher levels can cause extreme debilitation, loss of righting, coma, seizures and death. Long-term exposure has been shown to cause changes in neurophysiological or psychological capacity, and to increase the risk of lung, skin and bladder cancers, as well as having other carcinogenic effects (Hutcheson et al. 1996); (Boffetta et al. 1997); (Ritchie et al. 2001). Due to its toxicity, several environmental regulations have been enforced worldwide to protect the environment from PHCs contamination. As a result, attention has also been devoted to rehabilitating PHCs-contaminated soils. For both the diagnosis of suspected areas and the possibility of controlling the rehabilitation process, there is a great need to develop and implement a method that can rapidly detect small amounts of PHCs in soils as well as identify PHCs type, for better soil management and maintenance as a source of life and food.

4.1.1. Total Phcs

Various fuel products are the outcome of crude-oil refinement. These products contain mostly C (83-97%), H (10-14%), N (0.1-2%), O (0.0005-1.5%), S (0.05-6%) and various types of metals such as V, Fe, Cu and so on (<0.1%). These fuel products are classified by two main criteria: structure of the hydrocarbon (aliphatic or aromatic), and number of C in the chemical structure. These two criteria affect several of the products' characteristics, such as boiling point, solubility, absorption and toxicity. There are some overlapping characteristics among the various groups of hydrocarbons but in general, aromatic compounds are more toxic than aliphatic structures and there is a connection between the length of an aliphatic molecule and its toxicity (Tetra Tech EM Inc. for EPA 2000). Total PHCs (TPH) is a commonly used measurement index for quantifying environmental contamination originating from fuel. Choosing an analytical method for assessing TPH in soil samples is a problematic task due to the complex nature of the fuel ingredients and the fact that those ingredients change with time. Today, the most commonly used analytical method for assessing TPH in soil samples is based on the no-longer approved Environmental Protection Agency (EPA) method 418.1. This method was originally intended for assessment of TPH in water (EPA 1978), but was later adjusted to assess TPH in soil samples (EPA 1983). Not only has this previously approved method been withdrawn by the EPA, but it is also problematic for various other reasons.

4.1.2. Analytical Methods

4.1.2.1. Epa Method 418.1

The most common method in use today, which is based on the EPA's method 418.1 which was adjusted for soils, is costly and time-consuming. Briefly, the sample is brought to the laboratory and a specific extraction technique is applied using Freon 113 (trichlorotrifluoroethane). The TPH level is then determined in the solvent by Fourier transform infrared (FTIR) spectrophotometer. This method necessitates careful shipping of the samples to the laboratory (e.g., cooling conditions and use of special quartz tubes) and several replications to precisely establish the chemical analysis. Measuring one sample can cost about $50 and the results cannot provide information in a near-real-time domain. Consequently, in obtaining a reliable spatial pattern of the contaminated area, many samples are required to generate useful information that will facilitate the rehabilitation or precautionary processes. This method can only be used for quantification of TPH and not for identifying PHCs type.

4.1.2.2. Gravimetric

Hydrocarbons are removed from the sample by an extraction solvent, the solvent is weighed before and after the extraction, and the difference in weight expresses the TPH. This method is analogous to weighing a battleship with its captain, then subtracting the weight of the battleship in order to find out the weight of the captain: while it is relatively inexpensive, it is highly inaccurate.

4.1.2.3. Epa Method 8440

Hydrocarbons are extracted from the sample with CO_2 at very high pressure (340 atm), at 80°C for 30 min. After extraction, TPH is determined by FTIR spectrometry and compared to a calibration model (similar to method 418.1).

This method is not only expensive but it is also dangerous due to the high pressures involved.

4.1.3. Tph Field Methods

The abovementioned laboratory methods are costly and time-consuming, and usually cannot be implemented in a real-time domain at the contaminated site. Samples need to be transported to the laboratory in a cooled container and there is a waiting period before results are reported. In addition to the laboratory tests, a few field-measurement kits are available. The seven field-measurement devices listed below were tested at the Navy Base Ventura County site in Port Hueneme, California by Tetra Tech EM Inc. (2000):

- CHEMetrics, Inc.: RemediAid™ Total Petroleum Hydrocarbon Starter Kit
- Wilks Enterprise, Inc.: Infracal® TOG/TPH Analyzer, Models CVH and HATR-T
- Horiba Instruments, Inc,: OCMA-350 Oil Content Analyzer
- Dexsil® Corp.: PetroFLAG™ Hydrocarbon Test Kit for Soil
- Environmental Systems Corp.: Synchronous Scanning Luminoscope
- siteLAB® Corp.: Analytical Test Kit UVF-3100A
- Strategic Diagnostics, Inc.: EnSys Petro Test System

These field tests, making use of several technologies, are summarized in Table 2.

These technologies are thoroughly discussed in the report from the EPA field test conducted in June 2000: while some of these tests produced good results, the kits were found to be expensive and to require a knowledgeable operator (Tetra Tech EM Inc. for EPA 2000).

All analytical methods, whether conducted in the laboratory or *in situ*, that are available for the determination of TPH in soil samples have individual drawbacks, for example: shipping of the sample to a distant laboratory (no *in-situ* availability), long waiting periods for results, high cost per sample, inaccuracy, and the requirement of a skilled staff to perform the measurement. Due to these drawbacks, a new low-cost *in-situ* method for assessing PHC contamination, particularly at a new site, is greatly needed. Emphasized desired properties of this new method are cost-effectiveness and its ability to be implemented by not only highly trained personnel.

Table 2. Technology of TPH field-measurement kits

Technology	Measurement Device
Friedel-Crafts alkylation reaction and colorimetry	RemediAid™ Total Petroleum Hydrocarbon Starter Kit
Infrared analysis	Infracal® TOG/TPH Analyzer, Models CVH and HATR-T; OCMA-350 Oil Content Analyzer
Emulsion turbidimetry	PetroFLAG™ Hydrocarbon Test Kit for Soil
Ultraviolet fluorescence spectroscopy	Synchronous Scanning Luminoscope; siteLAB® Analytical Test Kit UVF-3100A
Immunoassay and colorimetry	EnSys Petro Test System

4.2. Tph And Reflectance Spectroscopy

4.2.1. Absorption Features Of Organic Compounds

Carbon-hydrogen bonds (C-H, C-H_2, C-H_3), hydroxy groups (O-H), double and triple bonds of aliphatics and aromatics, carboxyl groups (C=O), ester groups (C-O-C), amino groups (N-H) and other structural groups of organic chemicals exhibit characteristic fundamental vibrations that are evident in the 2,500 to 6,670 nm wavelength region of the IR spectrum. Furthermore, these compounds exhibit overtone and combination bands in the IR and VNIR-SWIR spectrum between 500 and 6,670 nm. These fundamental vibrations can be determined by IR spectroscopy and some of the overtone and combination bands can be determined by VNIR-SWIR spectroscopy (Winkelmann 2005).

4.2.2. Vnir-Swir Spectroscopy Of Organic Chemicals

As IR spectroscopy is the better choice for qualitative analysis, VNIR-SWIR spectroscopy is primarily applied for quantitative analysis in many disciplines (termed "chemometrics"). However, despite the fact that VNIR-SWIR spectra of organic chemicals are often highly complex due to many overlapping overtone and combination absorption features, they still contain information on the functional and structural groups of the chemicals under investigation. During the past 10 years, a new quantitative methodology has been widely developed that utilizes the reflected radiation in solids. Basically, this technology was adopted 40 years ago from a strategy developed in the food science discipline (Ben-Gera & Norris 1968a; Ben-Gera & Norris 1968b), whereas today it is widely utilized in many industrial and scientific applications. In this approach, the radiation reflected across the VNIR-SWIR region is modeled against constituents determined by traditional chemical analysis and then used to predict unknown samples. This technology permits rapid and cost-effective measurements on many samples at a given location and therefore significantly reduces chemical analysis costs. Despite the heterogeneity of the methodology, it was demonstrated by many (e.g: Nanni & Demattê 2006; Ben-Dor, et al. 2008; Rossel & Behrens 2010) that soil constituents can be extracted from a simple reflectance measurement (laboratory and field) if an appropriate calibration model has been previously developed. Since Ben-Dor et al. (1995)'s pioneering study, many other workers have explored this promising technology for soils and a significant breakthrough in this area was its replacement of wet chemistry in many scenarios. The adoption of this approach to evaluate hydrocarbon contamination in soil is obvious and a few workers have partially demonstrated this application, as described in the next section. As shown above, VNIR-SWIR spectroscopy holds great potential for quantitative chemical analysis. If calibration standards are used, the identification of known substances in a mixture and the determination of their concentrations are possible. Furthermore, it seems likely that the identification of substances based on their VNIR-SWIR spectra will become possible for classification purposes.

4.3. Reflectance Spectroscopy Of Phcs

The spectral properties of hydrocarbons were identified in the late 1980s, although it has been argued that these properties were only visible at concentrations of 4% wt (by weight) and above (Cloutis 1989). In the mid 1990s, a VNIR-SWIR reflectance sensor was developed as a proof of concept for the detection of OM in soil based on the same spectral properties identified by Cloutis. The sensor was designed for the detection of benzene in soil at a minimal concentration of 4.4% wt: several configurations were tested and minimal information was provided (Schneider et al. 1995). Soon after, the U.S. Department of Energy contracted a private company to investigate the application of reflectance spectroscopy to determine motor oil contamination in sandy loam. A schematic design for a field instrument was suggested, although only one contaminant and one type of soil were tested, using very few samples with a very limited range of contamination (Stallard et al. 1996). A more inclusive study was conducted shortly thereafter, using three types of soil contaminated in the laboratory with diesel and gasoline. A 0.1% wt and 0.5% wt minimum detection limit was achieved, respectively (Zwanziger & Heidrun 1998). The first study utilizing field-collected samples was not able to produce robust models, resulting in a very low correlation coefficient ($R^2 = 0.46$) and large errors, probably due to very low sample count and problems with the chemical measurements performed in the laboratory which produced inconsistent results (Malley et al. 1999). Attempts at mapping hydrocarbons using Landsat and Daedalus in 1994 and 1995 failed, probably due to the limited spectral resolution of the sensors (multispectral sensors); on the other hand, the higher spatial and spectral resolution, as well as the very high signal-to-noise ratio of the airborne hyperspectral sensor used (HyMap) (Cocks et al. 1998) yielded successful identification of hydrocarbon- and oil-contaminated soils, but only for high contaminant concentrations (2.5% wt) (Hörig et al. 2001). Based on the HyMap mission, a Hydrocarbon Index was developed for the mapping of hydrocarbon-bearing materials. This index is limited to very high signal-to-noise ratio sensors, as well as by other issues such as problems with land cover, vegetation and high concentration detection levels (Kühn et al. 2004). The most comprehensive work on reflectance properties of hydrocarbons was conducted by Winkelmann (2005): several types of hydrocarbons were mixed with several types of soil under laboratory conditions. They were measured spectrally and an attempt was made to separate them into hydrocarbon groups using the reflectance spectra; hyperspectral airborne remote sensing was also applied to identifying hydrocarbon contamination. No quantitative models were tested, although this was mentioned as an avenue of further study (Winkelmann 2005). A recent study by Chakraborty et al. (2010) on the prediction accuracy of VNIR-SWIR reflectance spectroscopy of petroleum contaminated soil, showed fair validation results (R^2=0.64). The study included 46 field collected samples that were preprocessed and modeled by several techniques.

While the abovementioned studies addressed only concentration levels >0.1%, today's environmental regulations require identification levels that are an order of magnitude lower. Moreover, none of the above studies dealt with identifying PHCs type and mixtures in soil using reflectance spectroscopy. A comprehensive study including several types of soil with several types of PHCs at a large variety of contamination levels has not yet been published. Most of the studies were performed with lab-prepared samples and almost none collected in the field. Furthermore, no generic models were developed, especially combining laboratory and field samples, and no real quantitative operation model was presented for real-life applications.

A long-term study conducted by Schwartz (2012) over the past 3 years included the development of NIRS models for the prediction of TPH in soil samples at high precision for low contamination levels, as well as PHCs classification. Five types of soil were mixed with three types of PHCs at 50 concentration levels, giving 750 laboratory samples, along with more than 200 other laboratory-prepared samples with various mixtures to study the soil properties' effects on spectral measurements and chemical analysis. The samples were evaluated for TPH level with EPA method 418.1, as well as measured spectrally by an ASD Fieldspec Pro. An "all possibilities" approach was taken for the creation of robust NIRS models, including the evaluation of many preprocessing techniques (smoothing, absorbance, first and second derivatives, and continuum removal), as well as PLS and ANN modeling methods. Forty-one soil samples were then collected in a field campaign and TPH level was predicted by both the PLS and ANN models, as well as being determined by the traditional EPA 418.1 method. A detection limit of 1,000 ppm was set, and samples with concentrations of 1,000 ppm or lower (27 samples) were successfully predicted as non-contaminated or with very low contamination. The rest of the predicted samples gave a correlation between the predicted and measured values of $R^2 > 0.9$ (this study will be published soon as a Ph.D. thesis, and as several papers in peer-reviewed journals).

5. ACID MINE DRAINAGE

Acid mine drainage (AMD) from mine wastes is considered a major problem in mining areas, because of the risk of heavy metal contamination of soils and water, Reflectance spectroscopy and NIRS is also used in the field of AMD. AMD is a byproduct of the oxidation of iron disulfide mineral (pyrite). While AMD reacts with the surrounding rocks and soil, heavy metals become more soluble and they can leach into the environment. This process is also substantially enhanced by bacterial action. In mining areas, the release of heavy metals is accelerated due to the increased oxidation rates, which are caused by the greater accessibility to air through mine workings, waste rock, and tailings by mineral processing. Furthermore, there is the added risk of mining accidents such as in Aznalcóllar (Spain 1998) or Baia Mare/Borsa (Romania 2000). During the course of these accidents, huge amounts of mine waste and toxic substances (heavy metals, cyanide) were set free in one go, contaminating rivers and alluvial soils (Kemper & Sommer 2002). Pyrite oxidation is one of the most acid-producing natural weathering processes, in which trace metals are mobilized and released into the river system and groundwater. It is a complex process that proceeds rapidly when pyrite is exposed to air. In a first step, pyrite produces a solution of ferrous sulfate and sulfuric acid. The dissolved ferrous iron continues to oxidize and hydrolyze, producing additional acidity. During the oxidation process, the pyrite transforms first to copiapite, then to jarosite, schwertmannite, ferryhydrite and eventually to hematite or goethite (Kemper & Sommer 2003).

Based on the sulfur content, other information can also be derived. Sulfur plays a key role in the assessment, because it is one of the two elements forming pyrite (FeS_2), which is the main component of the sludge (75-80% wt), and it is the main source of acidity when pyrite oxidizes. The sulfur content is also used in conventional mining waste analysis for the prediction of its geochemical behavior and to identify wastes that are likely to be acid-generating or susceptible to heavy metal leaching (Kemper & Sommer 2003).

6. PESTICIDES

Soil contamination by agricultural pesticides is a matter of increasing concern. Several authors have reported accurate predictions of pesticide sorption to soil using NIRS. Bengtsson et al. (2007) used NIRS to predict lindane and linuron sorption to soil samples. The pesticide-content predictions were established by PLS of the VNIR-SWIR spectra and sorption data obtained using the batch equilibrium technique. PLS could predict 85% of the soil sorption (Kd) for lindane and 84% for linuron, respectively. Later, Brunet et al. (2009) tested the feasibility of determining chlordecone contamination in soils. Chlordecone, also known as kepone, is a chlorinated polycyclic ketone pesticide. It is toxic, persistent, and induces cumulative and delayed toxicity. The regular method to measure chlordecone involves extraction with two solvents and quantification by gas chromatography–mass spectroscopy. A total of 236 samples were collected at 0–30 and 30–60 cm depths in Andosols, Nitisols and Ferralsols from all over the eastern part of Martinique (French West Indies). Chlordecone content was determined by wet chemistry and the reflectance spectra were recorded. Modified PLS (mPLS) regression was used to correlate spectral data to conventional values. While conventional measures and NIRS predictions correlated poorly at high chlordecone contents (>12 mg kg^{-1}), when the validation set was divided into four classes based on chlordecone content, NIRS prediction properly classified 80% of the samples (Brunet et al. 2009).

7. VEGETATION STRESS AS AN INDICATOR FOR SOIL CONTAMINATION

The concept of using plants as indicators for soil contamination has been tested for many years, for many good reasons: 1) inorganics in general do not exhibit characteristic absorption features in the VNIR-SWIR wavelength region; 2) the plant's root system is susceptible to contaminated material, and extracts and transports it to the aboveground plant parts, and 3) soil is far from always being bare; it is often covered with plants. Free metal ions in the soil solution and exchangeable ions of clays and OM are mostly available for plant uptake (Lasat 2000). For their development, plants require not only macronutrients (N, P, K, S, Ca, and Mg), but also essential micronutrients such as Fe, Zn, Mn, Ni, Cu, and Mo. Most of these micronutrients accumulate in the plant tissues for their metabolic needs, but they never exceed 10 ppm. Higher levels of these metals (and other heavy metals) in plant tissue might have phytotoxic effects, sometimes resulting in death (Winkelmann 2005). Vegetation stress caused by heavy metal toxicity can show different symptoms, depending on the heavy metal's type and concentration, and on the plant species' sensitivity. The phototoxic effects typically cause physiological, morphological and ecological changes, expressed in many different ways, from chlorosis (reduced production of chlorophyll and hence reduced photosynthetic activity), leaf wilting, needle retention, branch density reduction, defoliation, growth inhibition, flowering and fruiting changes, dwarf growth and gigantism, to changes in plant species distribution, the establishment of adapted species, and the dying-off of whole plants or communities (Winkelmann 2005). However, stress can be induced in vegetation by a large number of other factors, including water deficiency, poor soil drainage, poor soil aeration, soil salinity, weed competition, pest infestation, nutrient deficiency, or nutrient poisoning (Levitt 1980; Lichtenthaler 1996).

A large number of studies have reported on spectra of metal-stressed vegetation being clearly different from those of unstressed vegetation. These differences usually include decreases in both NDVI and RVI with increasing plant stress (Davids & Tyler 2003; Dunagan et al. 2007), signs of reduced biomass and a shift of the red edge position (REP) (Kooistra et al. 2004; Dunagan et al. 2007) or red edge slope (Zhou et al. 2010), signs of reduced photosynthetic activity due to chlorosis and decreased reflectance in the 700 nm to 2,500 nm wavelength region (Kooistra et al. 2003). Some heavy metals which are considered contaminants are also essential micronutrients for all vegetation species, in particular Cu and Zn. Therefore, in some cases, even positive effects in reflectance spectra can be observed when one of the elements under investigation is clearly a micronutrient with limited availability. Such effects include a red-shift of the red edge and decreased reflection in the VIS wavelength region indicating increased photosynthetic activity (Horler et al. 1980). As no vegetation stress symptoms and corresponding spectral characteristic could be related specifically to heavy metal stress or any other contaminant stress, the sources causing the observed stress need to be carefully identified and separated.

8. LIMITATIONS, OBSTACLES AND PROBLEMS

8.1. Laboratory

A wide range of factors can affect soil reflectance spectra in both laboratory and field domains. In the lab, different spectrometers, or even repeating a specific sample's measurements in the same spectrometer may result in variations. Such variations might include subtle or strong alterations in wavelength location, peak absorption shape or radiometric intensity. In addition to the instrumentation itself, internal electronic noise can affect the measurements and mechanical noise factors (e.g. homogeneity and purity of the white reference panel, or subtle movement when holding the fiber optic) can strongly affect their consistency. In soil samples, where very weak spectral features are monitored for chemometric purposes, these noise factors can alter the robust use of a selected spectral model for a wide range of spectrometers and users. Recently, Pimstein et al. (2011) examined the variation stemming from the above sources using three ASD Fieldspec Pro spectrometers and developed a standard protocol for laboratory spectral operation. They also suggested using the same internal standard worldwide in order to correct the spectra of any spectrometer to a selected master spectrometer. They demonstrated that chemometric analytical results are more coherent and precise. A global soil spectral library, not just as with minerals, has to be followed by chemical information. Recently, Viscarra Rossel (Rossel, Soil Spectroscopy Group 2009) initiated a spectral archive containing more than 10,000 spectra taken worldwide and is in the process of integrating them into this library with some major soil attributes measured in the alternative “wet chemistry” way.

8.2. Field

In the field, there are more uncertainties than in the laboratory, such as atmosphere attenuation, sun angle, aspect and slope of the sample area, large pixel sample, BRDF effects and most of all, soil crusting and sealing which can emerge on any soil surface. The latter prevent representation of the real soil body in the measurement. In the laboratory, the soil is crushed to pass a 2 mm sieves, and all stones and litter debris are consequently removed before the measurement. In the field, those materials are present, as are physical and biogenic crusts and possible dust contamination. Another problem that might emerge in field soil measurement is the adjacency effect and the mixed pixel. The first can occur when the pixel in question is surrounded by bright reflective targets. The mixed pixel problem occurs when a pixel is composed of several chromophores. It is simple when the mixture is binary and complex when it is not. The sun's angle, slope and aspect might not be factors in the field if artificial illumination is used in a standard procedure. A contact probe equipped with tungsten-halogen illumination can be used, with caution. This is because the soil surface measured by such an instrument may consist of a very narrow field of view which might not represent the soil in question, but rather debris, stones or even soil aggregation.

8.3. Airborne and Spaceborne

Soil reflectance can also be measured from aircrafts and satellites, using either a point spectroscopy (Karnieli et al. 2001) or IS (Ben-Dor et al. 2009) sensor. However, more difficulties arise during such measurements to extract the correct reflectance, such as: small integration time (less photons), strong atmosphere effects, large pixel size and varying quality of the sensor’s stability and sensitivity. Brook & Ben-Dor (2011) have recently developed a more moderate method (SVC—Supervised Vicarious Calibration) to standardize all sensors' radiometric readings, with the aim of deriving the optimal soil reflectance from the airborne IS sensor. This method uses artificial net targets with varying densities placed on a bright background area. It is easy to use and performs well. The method has been recently validated in a European campaign over southern France, using three different sensors simultaneously onboard two different airplanes. The preliminary results were relatively good as compared to the alternative ways of deriving the reflectance from the airborne sensor (unpublished data). The artificial target and the suggested method help assess the atmospheric attenuation, and minimize sensor instability while correcting for the systematic noise. Another limitation using airborne IS for soil is its high cost, low availability, and the complexity involved in processing the raw data into a final reflectance product (Ben-Dor et al. 2009). These factors actually prevent ordinary users from using this technology, classifying it as an exclusive method. Vegetation coverage (partial or total) is also a problem for deriving correct soil reflectance from afar when it is mixed in the sensor’s field of view.

As a further step, the retrieval of quantitative mapping using spectral parameters derived from hyperspectral images should take into account differences in the spectral and spatial resolution between ground spectrometer and image data (Choe et al. 2008).

8.4. General Issues

In all of the above domains, it must be remembered that reflectance spectroscopy is strongly affected by water content, particle size distribution and the measurement protocol. In addition, ways of calculating the reflectance from radiometric readings, i.e. against a white standard panel or using the radiance-to-irradiance ratio, may change the final product. Furthermore, the reflectance represents only the surface and cannot provide information on the soil profile (unless a penetrating fiber optic is used such as that described by (Ben-Dor et al. 2008). When obtaining a soil spectrum from a user, a meta-data set which characterizes all of these factors (moisture content, grain size and method of reflectance calculation) is strongly needed, because they have a significant impact on the final spectra's behavior.

8.5. Summary

Soil reflectance is an inherent property of the soil but many factors can affect its performance. Internal standards, a standard protocol and controlled conditions are a few of the things that can assist in sharing and comparing soil spectra (and chemometric models) worldwide (e.g. the Global Soil Library by Rossel and Soil Spectroscopy Group (2009)). It is obvious that uncertainties in the laboratory are smaller than those in the field, and the latter are smaller than those obtained from air or space. Use of radiometric data acquired from remote sensing domains to measure soil reflectance information should therefore be undertaken with caution.

9. FUTURE POTENTIAL OF REMOTE SENSING TECHNOLOGY FOR ASSESSING SOIL CONTAMINATION

Soil reflectance has become a very useful tool over the past 20 years in the laboratory, in the field and from air and space. As the sensitivity of portable field spectrometers increases, field soil spectroscopy is expected to become a basic tool for rapid point-by-point monitoring of the soil environment. The commercial development, operation and use of air- and spaceborne image spectrometers can provide near-laboratory-quality spectra of every pixel in an image and very soon, will permit remote sensing of soils with high standards. Information about soils from reflectance spectra in the VNIR and SWIR spectral regions represents almost all of the data passive solar sensors can provide. It is anticipated that the thermal IR regions will also become part of the soil spectral arena as they contain diagnostic information on some soil attributes that the reflective spectral region does not. The development of a thermal spectrometer (either point or imaging) will enable obtaining more spectral information with better accuracy. Another future insight is the capability to measure the soil profile's spectral distribution using a fiber optic assembly and small boreholes. Today, the major limitation of this technique is the fiber optics' length (as the fiber length increases, the signal across the SWIR region decreases—today, the length is limited to 1.5 m). Miniaturizing the spectrometer will enable placing it on the penetrating optical head, thus bypassing the limitation imposed by fiber length. This, in turn, will enable spectral measurement of any profile depth which, when combined with the NIRS model, might be able to provide *in-situ* soil contamination attributes of the soil profile. Combining the penetrating optical device information with the hyperspectral remote sensing data will also enable extracting the spatial distribution of the contaminant in question in 3D view. Data merging of soil reflectance information with GIS layers and other potential sensors that can be used simultaneously is also a key vision for the future. Time-series spectral measurements of large areas are also very important. Future IS from orbit will enable global coverage of every point on the globe with very good signal-to-noise ratio, such as the PRISMA (Labate et al. 2009),HyspIRI (Knox et al. 2010) and EnMAP (Stuffler et al. 2007) programs. This will enable monitoring soil surface changes in short- and long-term scenarios. Another future activity related to soil spectroscopy and soil contamination assessment is the development of better computing systems in which better models to derive highly accurate soil attributes will be achieved. To that end, the "all possibilities" approach in which all statistical and data-manipulation techniques can be applied automatically to a set of data is strongly required. Computing power and simple operation algorithms are key factors to that end.

10. GENERAL CONCLUSIONS AND SUMMARY

Soil reflectance spectroscopy can assess soil properties rapidly and quantitatively in both point and spatial domains. Direct and indirect soil properties, as well as soil contamination characteristics, can be extracted efficiently at low cost *in situ*. To that end, the VNIR-SWIR spectra must be preprocessed and modeled against reference data obtained by traditional methods. While some soil contaminants are featureless in the VNIR-SWIR region, their detection and quantification is possible, as they may be detected indirectly based on their association with other detectable materials. Although a wide range of factors can affect soil reflectance spectra in both laboratory and field domains thus strongly affecting the consistency of the resulting measurements, recent developments and proper protocols are allowing for more consistent and accurate results. While using NIRS to predict soil contaminants can save time, some cases involve the trade-off of reduced accuracy. Thus, the spectral assessment of soil samples cannot completely replace, but rather complements the classical chemical analysis in these and other cases. The benefits of using NIRS can result in the practicable processing of a large number of samples and savings in chemicals, lengthy tedious processes and manpower. In terms of spatial analysis, an airborne or spaceborne hyperspectral sensor can be useful for the screening of large areas and the reproduction of the spatial distribution patterns of contaminated soil areas. Nevertheless, the field of reflectance spectroscopy as a tool for monitoring contaminated soils still requires further study toward increased accuracy and the development of practical real-life applications.

REFERENCES

1. E. Ben-Dor, 2002Quantitative remote sensing of soil properties. Advances in Agronomy, 75, 173243

2. E. Ben-Dor, A. Banin, 1995Near-infrared analysis as a rapid method to simultaneously evaluate several soil properties. Soil Science Society of America Journal, 59(2), 364372

3. E. Ben-Dor, et al.2009Using Imaging Spectroscopy to study soil properties. Remote Sensing of Environment, 113, S38S55

4. E. Ben-Dor, D. Heller, A. Chudnovsky, 2008A Novel Method of Classifying Soil Profiles in the Field using Optical Means. Soil Sci Soc Am J, 72(4), 11131123

5. E. Ben-Dor, J. A. Irons, G. F. Epema, 1999Soil reflectance. In Manual of Remote Sensing : Remote Sensing for the Earth Sciences. New York: John Wiley & Sons Inc, 111189

6. E. Ben-Dor, R. G. Taylor, et al.2008Imaging spectrometry for soil applications. Advances in Agronomy, 97, 321392

7. Ben-Gera, K. H. Norris, 1968aDetermination of moisture content in soybeans by direct spectrophotometry. Isrealic Journal of Agricultural Research, 18, 124132

8. Ben-Gera, K. H. Norris, 1968bInfluence of fat concentration on the absorption spectrum of milk in the near-infrared region. Israel J. Agric. Res, 18(3), 117124

9. S. Bengtsson, T. Berglöf, H. Kylin, 2007Near Infrared Reflectance Spectroscopy as a Tool to Predict Pesticide Sorption in Soil. Bulletin of Environmental Contamination and Toxicology, 78(5), 295298

10. P. Boffetta, N. Jourenkova, P. Gustavsson, 1997Cancer risk from occupational and environmental exposure to polycyclic aromatic hydrocarbons. Cancer Causes and Control, 8(3), 444472

11. S. A. Bowers, R. J. Hanks, 1965Reflectance of radiant energy from soils. Soil Science, 100(2), 130139

12. Brook, E. Ben-Dor, 2011Supervised vicarious calibration (SVC) of hyperspectral remote-sensing data. Remote Sensing of Environment, 115(6), 15431555

13. D. J. Brown, et al.2006Global soil characterization with VNIR diffuse reflectance spectroscopy. Geoderma, 132(3-4), 273290

14. D. Brunet, et al.2009Determination of soil content in chlordecone (organochlorine pesticide) using near infrared reflectance spectroscopy (NIRS). Environmental Pollution, 157(11), 31203125

15. S. Chakraborty, et al.2010Rapid Identification of Oil-Contaminated Soils Using Visible Near-Infrared Diffuse Reflectance Spectroscopy. Journal of environmental quality, 39.

16. E. Choe, et al.2009Qualitative analysis and mapping of heavy metals in an abandoned Au-Ag mine area using NIR spectroscopy. Environmental geology, 58(3), 477482

17. E. Choe, et al.2008Mapping of heavy metal pollution in stream sediments using combined geochemistry, field spectroscopy, and hyperspectral remote sensing: A case study of the Rodalquilar mining area, SE Spain. Remote Sensing of Environment, 112(7), 32223233

18. Christl, R. Kretzschmar, 1999Competitive sorption of copper and lead at the oxide-water interface: Implications for surface site density. Geochimica et Cosmochimica Acta, 63(19-20), 29292938

19. R. N. Clark, others, 1999Spectroscopy of rocks and minerals, and principles of spectroscopy. Manual of remote sensing, 3, 358

20. R. N. Clark, T. L. Roush, 1984Reflectance spectroscopy: Quantitative analysis techniques for remote sensing applications. Journal of Geophysical Research, 89(B7), 63296340

21. E. A. Cloutis, 1989Spectral Reflectance Properties of Hydrocarbons-Remote-Sensing Implications. Science, 245(4914), 165168

22. T. Cocks, et al.1998The HyMapTM airborne hyperspectral sensor: The system, calibration and performance. In 3742

23. C. Davids, A. N. Tyler, 2003Detecting contamination-induced tree stress within the Chernobyl exclusion zone. Remote Sensing of Environment, 85(1), 3038

24. G. Dekker, et al.2001Imaging spectrometry of water. Imaging Spectrometry: Basic principles and prospective applications, IV, Kluwer Academic Publishers, Dordrecht, The Netherlands.

25. S. C. Dunagan, M. S. Gilmore, J. C. Varekamp, 2007Effects of mercury on visible/near-infrared reflectance spectra of mustard spinach plants (Brassica rapa P.). Environmental Pollution, 148(1), 301311

26. Energy Information Administration,2009World Petroleum Consumption. International Petroleum Monthly.

27. K. H. Esbensen, et al.2002Multivariate data analysis: in practice: an introduction to multivariate data analysis and experimental design, Multivariate Data Analysis.

28. B. Hörig, et al.2001HyMap hyperspectral remote sensing to detect hydrocarbons. International Journal of Remote Sensing, 22(8), 14131422

29. D. N. H. Horler, J. Barber, A. R. Barringer, 1980Effects of heavy metals on the absorbance and reflectance spectra of plants. International Journal of Remote Sensing, 1(2), 121136

30. G. R. Hunt, 1980Electromagnetic radiation: The communication link in remote sensing. In Remote Sensing in Geology. New York: John Wiley &Sons Inc, 702

31. M. S. Hutcheson, et al.1996Beyond TPH: Health-Based Evaluation of Petroleum Hydrocarbon Exposures. Regulatory Toxicology and Pharmacology, 24(1), 85101

32. J. R. Irons, R. A. Weismiller, G. W. Peterson, 1989Soil reflectance, in theory and application of optical remote sensing. Remote Sensing, 66106

33. T. Isaksson, T. Naes, 1988The effect of multiplicative scatter correction (MSC) and linearity improvement in NIR spectroscopy. Applied Spectroscopy, 42(7), 12731284

34. L. Järup, 2003Hazards of heavy metal contamination. British Medical Bulletin, 68(1), 167182

35. J. Ji, et al.2010Diffuse reflectance spectroscopy study of heavy metals in agricultural soils of the Changjiang River Delta, China. In 19th World Congress of Soil Science.

36. Karnieli, et al.2001AFRI-aerosol free vegetation index. Remote Sensing of Environment, 77(1), 1021

37. T. Kemper, S. Sommer, 2002Estimate of Heavy Metal Contamination in Soils after a Mining Accident Using Reflectance Spectroscopy. Environmental Science & Technology, 36(12), 27422747

38. T. Kemper, S. Sommer, 2003Mapping and monitoring of residual heavy metal contamination and acidification risk after the Aznalcóllar mining accident (Andalusia, Spain) using field and airborne hyperspectral data. In Proceedings, 3rd EARSeL Workshop on Imaging Spectroscopy EARSeL Secretariat, Paris. 333343

39. R. G. Knox, et al.2010The Hyperspectral Infrared Imager (HyspIRI) mission: A new capability for global ecological research and applications. In The 95th ESA Annual Meeting.

40. L. Kooistra, et al.2003A comparison of methods to relate grass reflectance to soil metal contamination. International Journal of Remote Sensing, 24(24), 49955010

41. L. Kooistra, et al.2004Exploring field vegetation reflectance as an indicator of soil contamination in river floodplains. Environmental Pollution, 127(2), 281290

42. F. Kühn, K. Oppermann, B. Hörig, 2004Hydrocarbon Index-an algorithm for hyperspectral detection of hydrocarbons. International Journal of Remote Sensing, 25(12), 24672473

43. D. Labate, et al.2009The PRISMA payload optomechanical design, a high performance instrument for a new hyperspectral mission. Acta Astronautica, 65(9-10), 14291436

44. M. M. Lasat, 2000Phytoextraction of metals from contaminated soil: a review of plant/soil/metal interaction and assessment of pertinent agronomic issues. Journal of Hazardous Substance Research, 2(5), 125

45. R. M. Levenson, D. L. Farkas, 1997Digital spectral imaging for histopathology and cytopathology (Proceedings Paper).

46. J. Levitt, 1980Responses of plants to environmental stresses. Volume II. Water, radiation, salt, and other stresses., Academic Press.

47. H. K. Lichtenthaler, 1996Vegetation stress: an introduction to the stress concept in plants. Journal of Plant Physiology, 148(1), 414

48. J. R. Long, V. G. Gregoriou, P. J. Gemperline, 1990Spectroscopic calibration and quantitation using artificial neural networks. Analytical Chemistry, 62(17), 17911797

49. D. F. Malley, P. C. Williams, 1997Use of Near-Infrared Reflectance Spectroscopy in Prediction of Heavy Metals in Freshwater Sediment by Their Association with Organic Matter. Environmental Science & Technology, 31(12), 34613467

50. D. F. Malley, K. N. Hunter, G. R. B. Webster, 1999Analysis of Diesel Fuel Contamination in Soils by Near-Infrared Reflectance Spectrometry and Solid Phase Microextraction-Gas Chromatography. Soil and Sediment Contamination, 8(4), 481489

51. T. Naes, T. Isaksson, B. Kowalski, 1990Locally weighted regression and scatter correction for near-infrared reflectance data. Analytical Chemistry, 62(7), 664673

52. T. Naes, et al.2002A user friendly guide to multivariate calibration and classification, NIR Publications.

53. M. R. Nanni, J. A. M. Demattê, 2006Spectral Reflectance Methodology in Comparison to Traditional Soil Analysis. Soil Science Society of America Journal, 70(2), 393

54. M. Nicolaï, et al.2007Nondestructive measurement of fruit and vegetable quality by means of NIR spectroscopy: A review. Postharvest Biology and Technology, 46(2), 99118

55. M. Pandit, G. M. Filippelli, L. Li, 2010Estimation of heavy-metal contamination in soil using reflectance spectroscopy and partial least-squares regression. International Journal of Remote Sensing, 31(15), 41114123

56. Pimstein, E. Ben-Dor, G. Notesco, 2011Performance of Three Identical Spectrometers in Retrieving Soil Reflectance under Laboratory Conditions. Soil Science Society of America Journal, 75(2), 746

57. H. Y. Ren, et al.2009Analysis of visible and near-infrared spectra of As-contaminated soil in croplands beside mines. Guang pu xue yu guang pu fen xi= Guang pu, 29(1), 114

58. G. D. Ritchie, et al.2001A REVIEW OF THE NEUROTOXICITY RISK OF SELECTED HYDROCARBON FUELS. Journal of Toxicology & Environmental Health: Part B, 4(3), 223312

59. R. V. R. Rossel, Spectroscopy. Soil, Group, 2009The Soil Spectroscopy Group and the development of a global soil spectral library. Available at:http://adsabs.harvard.edu/abs/2009EGUGA..1114021RAccessed April 20, 2011].

60. R. A. V. Rossel, T. Behrens, 2010Using data mining to model and interpret soil diffuse reflectance spectra. Geoderma, 158(1-2), 4654

61. Schneider, et al.1995Fiber-optic near-infrared reflectance sensor for detection of organics in soils. Photonics Technology Letters, IEEE, 7(1), 8789

62. G. Schwartz, 2012Reflectance spectroscopy as a rapid tool for qualitative mapping and classification of hydrocarbons soil contamination. Ph.D Thesis, Tel Aviv University. (in preparation)

63. G. Siebielec, et al.2004Near-and mid-infrared diffuse reflectance spectroscopy for measuring soil metal content.

64. G. Soenksen, Y. Garini, I. Bar-Am, 1996Multicolor FISH using a novel spectral bioimaging system. In Proceedings of SPIE. 303

65. L. Sparks, 1995Environmental soil chemistry. New York.

66. B. R. Stallard, M. J. Garcia, S. Kaushik, 1996Near-IR Reflectance Spectroscopy for the Determination of Motor Oil Contamination in Sandy Loam. Applied Spectroscopy, 50(3), 334338

67. Stark, K. Luchter, M. Margoshes, 1986Near-Infrared Analysis (NIRA): A Technology for Quantitative and Qualitative Analysis. Applied Spectroscopy Reviews, 22(4), 335

68. E. R. Stoner, et al.1980Atlas of soil reflectance properties. Research Bulletin, Agricultural Experiment Station, Purdue University, (962).

69. T. Stuffler, et al.2007The EnMAP hyperspectral imager-An advanced optical payload for future applications in Earth observation programmes. Acta Astronautica, 61(1-6), 115120

70. D. G. Taylor, C. M. Nenadic, J. V. Crable, 1970Infrared Spectra for Mineral Identification. American Industrial Hygiene Association Journal, 31(1), 100

71. Tetra Tech EM Inc. for USEPA,2000Demonstration Plan- Field Measurement Technologies for Total Petroleum Hydrocarbons in Soil. United States Environmental Protection Agency, EPA/600/R-01/060, 1245

72. United States Environmental Protection Agency (USEPA),1983Methods for Chemical Analysis of Water and Wastes. Government Printing Office: Washington, DC.

73. United States Environmental Protection Agency (USEPA),1978Test Method for Evaluating Total Recoverable Petroleum Hydrocarbon, Method 418.1 (Spectrophotometric, Infrared). Government Printing Office: Washington, DC.

74. Vane, A. F. H. Goetz, J. B. Wellman, 1984Airborne imaging spectrometer: A new tool for remote sensing. IEEE Transactions on Geoscience and Remote Sensing, 22(6), 546549

75. M. Vohland, C. Bossung, H. C. Fründ, 2009A spectroscopic approach to assess trace-heavy metal contents in contaminated floodplain soils via spectrally active soil components. Journal of Plant Nutrition and Soil Science, 172(2), 201209

76. K. H. Winkelmann, 2005On the applicability of imaging spectrometry for the detection and investigation of contaminated sites with particular consideration given to the detection of fuel hydrocarbon contaminants in soil. Unpublished PhD Thesis. Brandenburg University of Technology.

77. S. Wold, H. Martens, H. Wold, 1983The multivariate calibration problem in chemistry solved by the PLS method. Matrix Pencils, 286293

78. Y. Wu, et al.2007A mechanism study of reflectance spectroscopy for investigating heavy metals in soils. Soil Science Society of America Journal, 71(3), 918926

79. Y. Z. Wu, et al.2005Feasibility of Reflectance Spectroscopy for the Assessment of Soil Mercury Contamination. Environmental Science & Technology, 39(3), 873878

80. J. M. Zachara, J. C. Westall, 1999Chemical modeling of ion adsorption in soils. Soil physical chemistry, 2, 4795

81. Zhou, et al.2010Heavy Metal Stress to P. Tomentosa in Coal Mining Area and Its Spectral Feature. In Bioinformatics and Biomedical Engineering (iCBBE), 2010 4th International Conference on. 14

82. Z. Zwanziger, F. Heidrun, 1998Near infrared spectroscopy of fuel contaminated sand and soil. I. preliminary results and calibration study. Journal of Near Infrared Spectroscopy, 6(1), 189197

Chapter 10

Heavy Metal Contamination in Soils of Remnant Natural and Plantation Forests in an Urbanized Region of the Pearl River Delta, China

Enqing Hou [1,2,†], Huimin Xiang [1,3,†], Jianli Li [4], Jiong Li [1,2] and Dazhi Wen [1,2,*]

[1]Institute of Ecology and Environmental Sciences, South China Botanical Garden, Chinese Academy of Sciences, Guangzhou 510650, China

[2]Key Laboratory of Vegetation Restoration and Management of Degraded Ecosystems, South China Botanical Garden, Chinese Academy of Sciences, Guangzhou 510650, China

[3]University of the Chinese Academy of Sciences, Beijing 100049, China

[4]Agricultural and Forestry Extension Centre of Nanhai District, Guangdong 528222, China

ABSTRACT

Remnant forests in urban areas provide vital ecosystem services but are susceptible to many human activities including heavy metal emissions. In this study, we collected 192 samples of mineral soils at depths of 0–3, 3–13 and 13–23 cm in 16 remnant forests (eight natural forests and eight plantation forests) in the urbanized Pearl River Delta, China. We assessed the potential risks of soil Cu, Zn, Pb, Mn, Ni and Cr to the vegetation in these forests based on their total and 0.1 M HCl extractable concentrations. The mean concentrations for all soil samples were 202.7, 102.0, 75.7, 24.3, 30.3, and 7.8 mg/kg for Zn, Mn, Pb, Cu, Cr, and Ni, respectively. Compared to background values, total soil Zn concentrations were higher for both the natural and plantation forests located near both industrial and non-industrial sites; total soil Cu and Pb concentrations were higher near industrial sites, particularly for the natural forests. Total soil Pb, Cu, and Mn concentrations and exchangeable soil Pb and Mn concentrations were higher in the natural forests than in the plantation forests. Total soil Cu and Pb concentrations and extractable soil Cu, Pb, Zn, and Mn concentrations decreased with soil depth. Based on these results and previous findings of continued acidification and low phosphorus availability of these soils, we recommend that the growth of these remnant forests can be improved by the application of phosphate rock.

KEYWORDS

heavy metals; soil; remnant forests; urbanization; Pearl River Delta

1. INTRODUCTION

Remnants of forest ecosystems occur in urban areas around the world [1,2]. Although forest remnants are typically small patches, they can provide important ecological services that improve human well-being in urban areas. They can, for example, improve urban air quality, ameliorate the heat island effect, preserve biodiversity, and provide wood products, recreational venues, and landscape aesthetics [1,2,3,4]. For these reasons, multidisciplinary studies have increasingly focused on the sustainability and conservation of these forests [1,2,3,5,6].

Relative to natural forests in rural areas, remnant forests in urban areas are typically more influenced by human activities, such as the building of houses and roads, recreation, and emissions of SO_x, NO_x, and particulate matters [2,7,8]. Of these activities, the emissions of pollutants are likely to influence forest ecosystems at a large spatial scale in urban and rural areas because of atmospheric transport [9,10,11]. The particulate matters emitted from industries and as a consequence of urban heating and other human activities usually contain high levels of heavy metals such as Pb and Zn [7,8,12]. A long-term continuous emission of particulate matters may result in the deposition of a significant amount of heavy metals in forest ecosystems, which may reduce biomass production, litter decomposition, and other ecosystem functions, especially in urban areas [9,13,14].

The Pearl River Delta (PRD) in southern China (Figure 1) has an area of 41,600 km^2 and had a population of over 56 million in 2010 [15]. The PRD has contributed greatly to the rapid economic development of China and is now the economic center of southern China. With rapid urbanization and industrialization, however, large areas of natural lands in the PRD have been converted into farmlands or urban areas [16,17]. At the same time, various harmful substances, including acid precursors and heavy metals, have been emitted into the environment [7,17,18] and have increasingly impacted the remnant forests in this area [19,20,21]. Previous studies reported that heavy metals in forest soils were generally higher in the urbanized (or industrialized) areas than in the rural (or non-industrialized) areas of the PRD [20,22]. However, it is still unclear how heavy metals in soils of remnant forests in the PRD and in other urbanized areas vary with site location, forest type, and soil depth.

In this study, we determined the total and 0.1 M HCl extractable concentrations of Cu, Zn, Pb, Mn, Ni, and Cr in mineral soils (0–3, 3–13, and 13–23 cm depths) in 16 patches of remnant forests (eight natural forests and eight plantation forests) in the PRD, China. Our purposes were: (1) to compare the distribution of heavy metals in soils of the remnant forests in industrial *vs.* non-industrial locations, in natural forests *vs.* plantation forests, and at three soil depths; and (2) to assess the potential risks and possible sources of these heavy metals in the studied forests.

2. MATERIALS AND METHODS

2.1. Site Description

This study was carried out in Foshan (22°38′–23°34′ N, 112°22′–113°23′ E), which is centrally located in the PRD of Guangdong Province (Figure 1). Foshan was one of the first three wholly urbanized, prefectural-level cities in China [23]. This region is characterized by a typical subtropical monsoon climate, with annual precipitation ranging from 1600 to 2000 mm and annual temperature ranging from 20 to 25 °C [17]. The forest soils are mainly Ferralsols [24], which developed from granite and sand shales.

Other than several large natural reserves in the suburban areas, remnant natural forests and remnant plantation forests are the major kinds of forest sites in the urban and suburban areas of the PRD [5,25], Remnant natural forests in the study area are also called "Fengshui Forests" or "Fengshui Woods" [3,25] and have been conserved as cultural heritage sites over the past several hundred years because of their socio-cultural significance for the indigenous people [26]. These forests are mainly near villages or graveyards and are generally dominated by native broadleaved evergreen species with community characteristics similar to those of regional climax forests [3,25]. Plantation forests with two rotations of reforestation are another major type of forests that remain in the area [25]. The first rotation was planted with Masson pine trees and/or eucalyptus species during the 1970–1980s, after the natural forests were harvested in 1950s. The second rotation was planted with native broadleaved tree species that corresponded to the tree species and community composition of the remnant natural forests after timber harvest of the Masson pine and/or Eucalyptus stands during 1997–2003 [25].

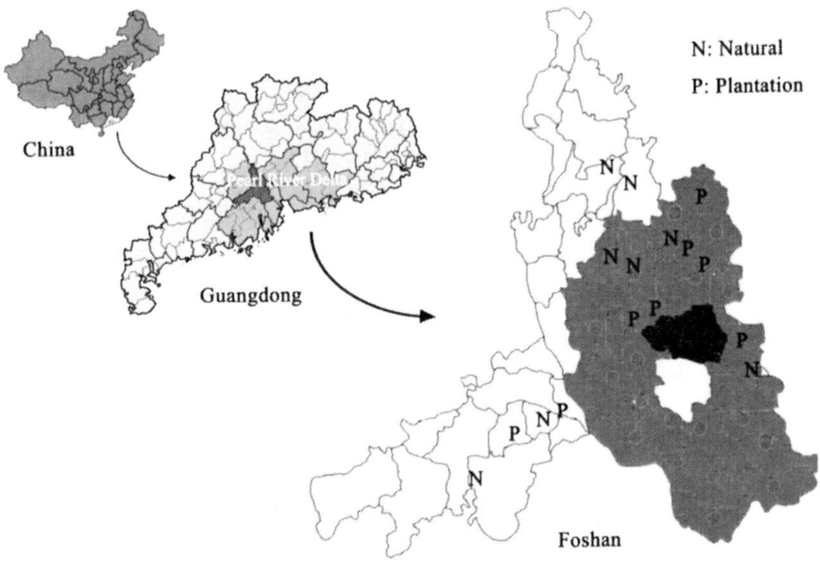

Figure 1. Location of 16 selected forest patches in Foshan, China. Black, grey and white areas in Foshan indicate the core urban, industrialized and non-industrial areas, respectively. "N" and "P" indicate the locations of remnant natural forests and plantation forests, respectively.

Eight patches of remnant natural forest and eight patches of remnant plantation forests in Foshan, China, were randomly selected (Figure 1). Site characteristics of the 16 remnant forest patches are shown in Table 1. Each site was categorized as an industrial or non-industrial location, according to the percentage of the local town's gross domestic product represented by gross industrial production (industrial towns, >60%; non-industrial towns, ≤60%) [17,27]. Because more remnant plantation forests in Foshan are located in industrial rather than in non-industrial locations, a simple random selection of eight remnant plantation forests resulted in six remnant plantation forests at industrial locations and two at non-industrial locations.

2.2. Sampling and Analysis

All soils were sampled during December 2008–January 2009. At each site, four sample locations that were >50 m apart were randomly selected. These areas lacked visible disturbance of vegetation, forest floor, or vertical soil profile. At each sample area, a soil pit of ~30 cm depth was excavated. After the litter and fermentation layers (usually 1–3 cm thickness) were removed, the mineral soil adjacent to the pit was collected at 0–3 cm, 3–13 cm, and 13–23 cm depths with a small stainless steel shovel. A 0–3 cm layer of mineral soil was separated because there was a color change at ~3 cm depth in the mineral soils at the 16 sites; this 0–3 cm layer may be defined as the Ah horizon according to the Canadian system of soil classification [28]. The 192 soil samples (16 sites × four areas within each site × three soil depths) were kept separate for analysis. After roots and stones had been removed, each sample was air-dried and then ground with a wooden stick until 100% of it passed through a 2-mm non-metal mesh. After the sample was mixed, a subsample (about 10 g) was oven-dried at 105 °C for 24 h, and a dry weight conversion was computed. Another subsample (about 30 g) was ground in an agate crucible until 100% of it passed through a 0.15-mm non-metal mesh; this sample was stored in a clean plastic bottle for the determination of total heavy metal concentrations. The remainder of the sample (>50 g) was stored in another clean plastic bottle for the determination of extractable heavy metal concentrations.

Total concentrations of heavy metals (Cu, Zn, Mn, Pb, Cr, and Ni) in soils were determined using an acid digestion method at atmospheric pressure [29,30]. Extractable concentrations of heavy metals in soils were determined using an acid extraction method [29]. A 10.00-g quantity of air-dried soil was placed in an acid-cleaned 150-mL Erlenmeyer flask, and 50 ml of 0.1 M analytical grade HCl was added. The flask was sealed with a rubber stopper and then shaken for 1.5 h at room temperature on a horizontal shaker (150 rpm) [29]. The mixture was filtered, and the concentrations of heavy metals in the solution were determined by ICP/AES (Optima 2000, Norwalk, CT, USA). Data quality was controlled by the use of national standard reference material (ESS-3) supplied by China National Environmental Monitoring Center. The results met the accuracy demand of Technical Specification for Soil Environmental Monitoring HJ/T 166-2004 [29]. The quantity of extractable heavy metal was expressed as a concentration (mg/kg) and as a percentage of the total heavy metal concentration. All results were expressed on an oven-dried mass.

2.3. Data Analyses

A three-way ANOVA (location × forest type × soil depth) using log-transformed data was carried out to determine the statistical significance of location type, forest type, soil depth, and their interactions. Significant interactions were assessed with one-way ANOVAs using Tamhane's *post hoc* test. The values obtained in the current study were compared to "background" values, which were published by the China National Environmental Monitoring Centre [31]; the samples (n = 167) used to obtain the background values were generally collected at 20 cm depth (the A horizon) in Guangdong Province and were collected and processed using the same general procedures as used in the current study. Correlations between soil heavy metal concentrations were investigated using Spearman rank test. Principal component analysis (PCA) was performed using the rotation method of Varimax with Kaiser Normalization, which allows identification of groups of metals with similar behaviors and common origins [14]. All analyses were performed using SPSS (version 16.0 for Windows, IBM, Chicago, IL, USA).

3. RESULTS

3.1. General Characteristics of Soil Heavy Metal Concentrations

Both the total and extractable concentrations of the six heavy metals varied greatly among the 192 soil samples. The coefficients of variation were larger for the extractable concentrations than for the total concentrations and were higher for Zn, Mn, and Pb than for Cu, Ni, and Cr (Table 2). The mean total soil concentrations of Zn, Pb, and Cu were 4.3-, 2.1-, and 1.4-times greater, respectively, than their background values, while the mean total concentrations for Cr, Ni, and Mn were lower than their background values (Table 2). Although the mean total Mn concentration in soil was lower than its background value, the upper limit was six times higher (Table 2). Therefore, our further analyses focused on Zn, Pb, Cu, and Mn concentrations in soil.

Table 1. Site characteristics of 16 selected forest patches in Foshan, China.

Site	Location Type	Forest Type	Grid Reference	Area (ha)	Major Tree Species
Kengmei	Industrial	Natural	112°58′ 29″ E, 23°10′ 11″ N	4.6	*Schefflera octophylla, Desmos chinensis, Alocasia macrorhiza*
Linyue	Industrial	Natural	113°14′ 02″ E, 23°00′ 13″ N	20.8	*Syzygium hancei, Symplocos lancifolia, Desmos chinensis*
Wanshi	Industrial	Natural	113°05′ 09″ E, 23°10′ 46″ N	6.9	*Engelhardtia roxburghiana, Aporosa dioica, Alchornea trewioides*
Yangao	Industrial	Natural	113°00′ 09″ E, 23°09′ 14″ N	3.0	*Syzygium hancei, Bambusa stenostachya, Ardisia hanceana*
Lunyong	Non-industrial	Natural	112°50′ 47″ E, 22°52′ 31″ N	9.5	*Schima superb, Indocalamus tessellates, Alpinia chinensis*
Shanbu	Non-industrial	Natural	113°00′ 07″ E, 23°16′ 36″ N	9.0	*Phoebe namu, Ardisia hanceana, Ixora chinensis*
Shukeng	Non-industrial	Natural	112°43′ 10″ E, 22°48′ 15″ N	5.9	*Machilus chinesis, Lasianthus chinensis, Indocalamus tessellates*
Yuantou	Non-industrial	Natural	112°56′ 07″ E, 23°18′ 16″ N	2.9	*Helicia cochinchinensis, Desmos chinensis, Ardisia hanceana*
Longtou	Industrial	Plantation	113°06′ 37″ E, 23°10′ 02″ N	5.6	*Cinnamomum camphora, Albizia falcataria, Delonix regia*
Sanguigang	Industrial	Plantation	113°00′ 17″ E, 23°03′ 46″ N	42.2	*Ficus altissima, Polyspora axillaris, Bombax malabaricum*
Sanshan	Industrial	Plantation	113°13′ 36″ E, 23°02′ 05″ N	32.8	*Ficus altissima, Cinnamomum camphora, Schima superba*
Shitang	Industrial	Plantation	113°04′ 25″ E, 23°16′ 18″ N	74.0	*Cinnamomum camphora, Castanopsis hystrix, Liquidambar formosana*
Xinjing	Industrial	Plantation	112°59′ 32″ E, 23°04′ 35″ N	104	*Ficus altissima, Ficus microcarpa, Bischofia javanica*
Zhanqigang	Industrial	Plantation	113°08′ 45″ E, 23°09′ 34″ N	83.7	*Cinnamomum camphora, Liquidambar formosana, Cinnamomum burmannii*
Xialiang	Non-industrial	Plantation	112°49′ 25″ E, 22°51′ 45″ N	10.7	*Cinnamomum camphora, Castanopsis hystrix, Liquidambar formosana*
Xian	Non-industrial	Plantation	112°52′ 37″ E, 22°52′ 33″ N	38.7	*Schima superb, Liquidambar formosana, Ficus microcarpa*

3.2. Comparison between Location Types and Forest Types and among Soil Depths

Total soil Cu, Mn, and Pb concentrations and extractable soil Cu, Zn, Mn, and Pb concentrations were all significantly higher in the industrial locations than in the non-industrial locations, while the trend for total soil Zn concentration was the opposite (Table 3 and Table 4). Total soil Cu, Mn, and Pb concentrations and extractable soil Mn and Pb concentrations were significantly lower in the plantation forests than in the natural forests (Table 3 and Table 4). In contrast, extractable soil Cu and Zn concentrations were significantly higher in the plantation forests than in the natural forests (Table 3 and Table 4), probably because more plantation forests than natural forests were in industrial locations (Table 4). For all soils, total soil Cu and Pb concentrations and extractable soil Cu, Pb, Zn, and Mn concentrations significantly decreased with soil depth (Table 3 and Table 4).

For total soil Pb concentration and extractable soil Mn concentration, interactions between location type and forest type were significant because values were relatively higher in the industrial locations than in the non-industrial locations particularly for the natural forests (Table 3 and Figure 2). For total and extractable soil Cu concentrations, interactions between location type and soil depth were significant because values were relatively higher in the industrial locations than in the non-industrial locations particularly for the upper layer of soil (Table 3 and Figure 3).

Total soil Cu and Pb concentrations were greater than the background concentrations in the industrial locations but were similar to the background concentrations in the non-industrial locations (Table 4). Total soil Zn concentration was markedly higher than the background concentration for all soil depths, both forest types, and both location types (Table 4). The mean total soil Mn concentration was much lower than its background value for both location types and forest types except for the natural forests in the industrial locations where the mean total soil Mn concentration was similar to the background concentration (Table 4).

The concentration of each extractable heavy metal expressed as a percentage of the total concentration of heavy metals and across all soils was highest for Pb (mean 28.1%) and Cu (25.8%); was intermediate for Mn (10.3%), Ni (7.8%), and Zn (5.8%); and was lowest for Cr (1.0%) (Table 5). The percentages of extractable Zn, Mn, Pb, and Ni were significantly higher in the industrial than in the non-industrial locations (Table 5).

Table 2. General characteristics of total and extractable concentrations (mg/kg) of six heavy metals in soils in 16 remnant forests in Foshan, China.

Parameter	Range	Median	Mean	SD	CV (%)	Background [§]
Total Cu	2.2–142.8	15.7	24.3	21.0	86.3	17.0
Total Zn	25.3–999.0	111.9	202.7	206.6	101.9	47.3
Total Mn	7.5–1452.5	32.2	102.0	257.2	252.2	279.0
Total Pb	1.7–784.8	25.2	75.7	144.9	191.3	36.0
Total Ni	0.7–34.4	7.0	7.8	4.7	60.0	14.4
Total Cr	1.0–129.2	27.0	30.3	18.2	59.9	50.5
Extractable Cu	0.2–51.6	3.3	4.6	5.4	118.6	
Extractable Zn	N.D.–101.8	2.8	6.4	10.8	169.1	
Extractable Mn	0.1–535.0	1.8	19.7	69.2	351.4	
Extractable Pb	N.D.–383.0	5.5	25.0	63.0	252.3	
Extractable Ni	N.D.–6.4	0.2	0.5	0.8	165.6	
Extractable Cr	N.D.–1.5	0.1	0.2	0.2	101.5	

n = 192 for values generated in the current study; [§] Background value of soils in the A horizon (generally about 20 cm depth) in Guangdong Province where Foshan is located. Values are means (n = 167) published by the China National Environmental Monitoring Centre [31], and these values were generally obtained using methods of field sampling, sample processing, and digestion that were similar to those used in the present study; N.D. indicates not detectable.

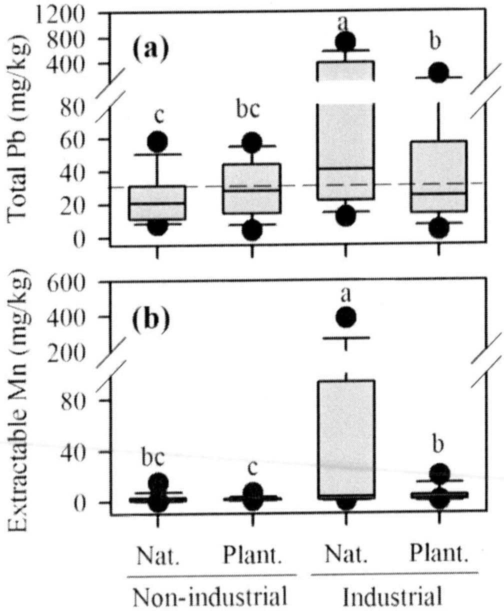

Figure 2. Total Pb and extractable Mn concentrations in soils from different location types (industrial *vs.* non-industrial) and forest types (Nat.: Natural forest; Plant.: Plantation forest). Whiskers indicate the 10th and 90th percentiles; the line in the box indicates the median; dots indicate the 5th and 95th percentiles of the outliers. Different letters indicate significant differences ($p < 0.05$) between soil groups. Dotted line indicates the background value of Guangdong province [31]. Soil total Pb (**a**) and extractable Mn (**b**) concentrations were relatively higher in the industrial locations than in the non-industrial locations particularly for the natural forests.

3.3. Correlations and PCA Analysis

Total soil Cu, Zn, and Mn concentrations were positively correlated with each other but the relationships were weak (r =0.18–0.33, $p < 0.05$; Table 6). Extractable soil Cu, Zn, Mn, Pb, and Ni concentrations were positively correlated with each other (r = 0.32–0.71, $p < 0.05$; Table 6). Positive correlations were also evident between the total and the extractable soil concentrations for Cu (r = 0.32, $p < 0.05$), Mn (r = 0.47, $p < 0.05$), and Pb (r = 0.64, $p < 0.05$). According to PCA analysis, PC1 explained 32.5% of the total variation and was occupied by total soil Cu, Pb, and Mn concentrations and extractable soil Mn and Pb concentrations (Figure 4); PC2 explained 18.3% of the total variation and was occupied by extractable soil Zn and Cu concentrations (Figure 4).

Table 3. Three-way ANOVA statistics of location type, forest type, soil depth, and interactions on total and extractable concentrations of Cu, Zn, Mn and Pb in soils in 16 remnant forests in Foshan, China.

Parameters	Values of F and *P* Statistics for the Independent Variables							
	Statistic	Location type (L)	Forest Type (F)	Soil Depth (D)	L × F	L × D	F × D	L × F × D
df		1	1	2	1	2	2	2
Total Cu	F	31.8	15.0	4.5	0.5	3.7	0.4	0.1
	P	0.000	0.000	0.012	0.466	0.025	0.648	0.883
Total Zn	F	7.8	1.3	0.6	9.3	0.3	0.1	0.1
	P	0.006	0.250	0.557	0.003	0.728	0.918	0.945
Total Mn	F	14.5	38.6	0.3	1.9	0.7	0.1	0.1
	P	0.000	0.000	0.707	0.169	0.513	0.916	0.952
Total Pb	F	18.9	4.7	6.7	10.6	1.4	0.1	0.3
	P	0.000	0.032	0.002	0.001	0.247	0.886	0.759
Extractable Cu	F	89.0	4.0	43.7	4.1	5.5	0.6	0.2
	P	0.000	0.048	0.000	0.043	0.005	0.532	0.780
Extractable Zn	F	53.4	5.6	77.4	3.9	3.0	1.7	0.4
	P	0.000	0.019	0.000	0.051	0.054	0.179	0.665
Extractable Mn	F	15.1	7.8	13.6	4.6	0.9	0.1	0.6
	P	0.000	0.006	0.000	0.033	0.421	0.897	0.538
Extractable Pb	F	26.9	6.5	19.4	0.2	1.3	0.9	0.5
	P	0.000	0.012	0.000	0.669	0.273	0.420	0.629

P values < 0.05 are in bold font.

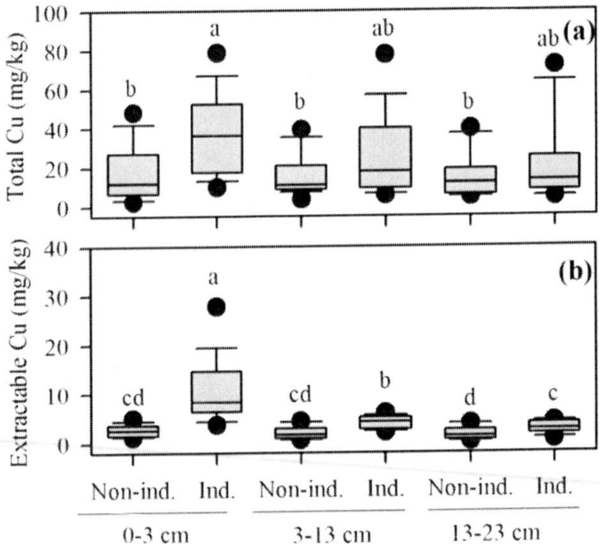

Figure 3. Total (**a**) and extractable (**b**) Cu concentrations in soils as affected by soil depth and location type (Non-ind.: non-industrial; Ind.: industrial). Whiskers indicate the 10th and 90th percentiles; the line in the box indicates the median; dots indicate the 5th and 95th percentiles of the outliers. Different letters indicate significant differences ($p < 0.05$) between soil groups. Soil total (**a**) and extractable (**b**) Cu concentrations were relatively higher in the industrial locations than in the non-industrial locations particularly for the upper layer of soil.

4. DISCUSSION

4.1. Potential Risks of Heavy Metals in the Soils at Industrial and non-Industrial Locations

Total soil Cu, Pb, and Zn concentrations in the industrial locations were all considerably higher than the background values. Total soil Cu and Pb concentrations in the industrial locations were also significantly higher than those in the non-industrial locations. These results were consistent with a previous study on the concentrations of these metals in plant and soil samples in large forest reserves along an urban-rural gradient in the PRD [22]. The results of the previous and current study indicate that forest ecosystems in the urbanized areas of the PRD general suffer from Cu, Pb, and Zn contamination.

Table 4. Mean values of total and extractable metal concentrations in soils as affected by soil depth, forest type and location type.

Soil Depth (cm)	Forest Type	Location Type	*n*	Cu	Zn	Mn	Pb	Ni	Cr
Total Concentration (mg/kg)									
0–3	Natural	Non-industrial	16	22.0	334.1	47.9	28.1	10.3	27.5
		Industrial	16	48.9	129.8	304.5	256.2	8.1	30.5
	Plantation	Non-industrial	8	10.2	121.9	15.9	34.1	12.9	79.8
		Industrial	24	33.2	171.1	47.4	65.3	5.5	25.5
3–13	Natural	Non-industrial	16	18.3	331.3	50.8	21.8	9.0	23.4
		Industrial	16	30.7	96.4	273.3	134.5	8.4	30.1
	Plantation	Non-industrial	8	10.1	124.8	13.9	25.0	12.2	53.3
		Industrial	24	21.8	195.1	41.2	53.7	5.0	22.4
13–23	Natural	Non-industrial	16	17.7	344.4	62.5	25.8	10.2	26.5
		Industrial	16	25.8	105.7	273.7	148.3	7.1	30.4
	Plantation	Non-industrial	8	9.3	159.8	16.9	28.5	10.1	52.9
		Industrial	24	20.8	225.4	36.6	48.0	4.6	20.3
	Background $		167	17.0	47.3	279.0	36.0	14.4	50.5

Extractable Concentration (mg/kg)

0–3	Natural	Non-industrial	16	2.4	4.0	4.2	6.5	1.1	0.3
		Industrial	16	14.7	19.6	92.0	93.2	0.9	0.3
	Plantation	Non-industrial	8	3.3	7.7	2.8	9.1	0.5	0.2
		Industrial	24	9.0	18.5	11.4	25.2	1.2	0.2
3–13	Natural	Non-industrial	16	1.8	1.5	2.1	4.3	0.2	0.3
		Industrial	16	3.9	4.9	54.3	63.8	0.3	0.1
	Plantation	Non-industrial	8	2.4	1.9	1.0	2.1	0.2	0.2
		Industrial	24	4.1	4.0	2.4	7.7	0.3	0.1
13–23	Natural	Non-industrial	16	1.6	1.2	3.6	3.9	0.1	0.3
		Industrial	16	2.4	2.9	55.1	64.2	0.2	0.1
	Plantation	Non-industrial	8	1.9	1.9	0.9	2.7	0.1	0.2
		Industrial	24	2.9	2.0	1.2	5.0	0.1	0.1

§ Background value of soils in the A horizon (generally about 20 cm depth) in Guangdong province, China where Foshan located, data is mean value, n = 167 [31].

Table 5. The concentration of each extractable heavy metal expressed as a percentage of the total concentration of heavy metals as affected soil depth, forest type, and location type.

Soil Depth	Forest Type	Location Type	*n*	Cu	Zn	Mn	Pb	Ni	Cr
0–3 cm	Natural	Non-industrial	16	25.1	2.4	9.8	31.7	10.8	1.0
		Industrial	16	29.3	14.8	20.7	28.3	11.3	1.3
	Plantation	Non-industrial	8	40.0	6.1	18.1	29.0	4.4	0.3
		Industrial	24	37.4	17.5	21.4	49.3	24.2	0.7
3–13 cm	Natural	Non-industrial	16	17.0	1.1	4.1	25.1	2.5	1.5
		Industrial	16	18.6	5.7	11.2	28.4	3.7	0.6
	Plantation	Non-industrial	8	23.3	1.8	7.7	12.6	1.8	0.4
		Industrial	24	30.6	5.5	6.3	30.7	5.9	0.7
13–23 cm	Natural	Non-industrial	16	20.7	0.9	4.2	23.6	6.4	1.3
		Industrial	16	18.9	2.9	10.1	21.1	2.8	0.5
	Plantation	Non-industrial	8	22.3	1.3	5.7	12.9	1.2	0.6
		Industrial	24	23.3	2.2	4.4	21.6	5.0	1.9

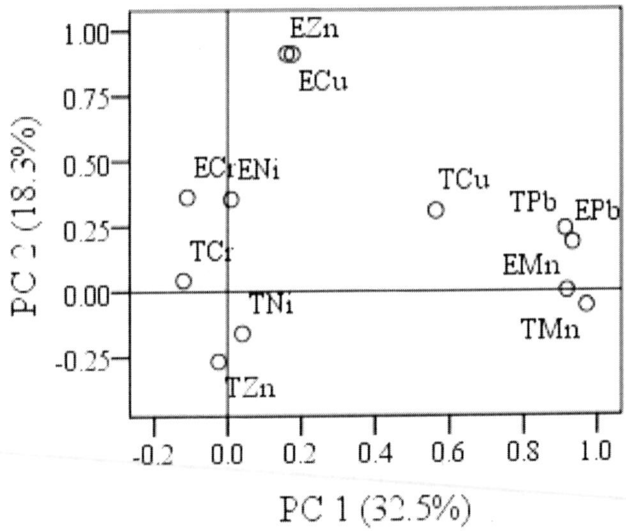

Figure 4. Principal component analysis (PCA) of total and extractable concentrations of six heavy metals in soils in 16 remnant forests in Foshan, China. TCu, TZn, TMn, TPb, TCr and TNi indicate total soil concentrations of Cu, Zn, Mn, Pb, Cr and Ni, respectively; ECu, EZn, EMn, EPb, ECr and ENi indicate extractable soil concentrations of Cu, Zn, Mn, Pb, Cr and Ni, respectively.

Table 6. Coefficients (r values) of correlations among total and extractable metal concentrations in soils in 16 remnant forests in Foshan, China.

	Total Cu	Total Zn	Total Mn	Total Pb	Total Ni	Total Cr	Ex. Cu	Ex. Zn	Ex. Mn	Ex. Pb	Ex. Ni
Total Zn	**0.28**										
Total Mn	**0.33**	**0.18**									
Total Pb	**0.26**	**−0.20**	**0.16**								
Total Ni	0.08	**0.23**	−0.04	0.05							
Total Cr	−0.11	0.10	**−0.39**	0.04	**0.71**						
Ex. Cu	**0.32**	**−0.25**	0.01	**0.36**	**−0.28**	−0.06					
Ex. Zn	**0.39**	−0.04	0.13	**0.39**	−0.09	−0.07	**0.59**				
Ex. Mn	**0.22**	**−0.19**	**0.47**	**0.49**	−0.09	**−0.16**	**0.50**	**0.61**			
Ex. Pb	**0.36**	−0.08	**0.37**	**0.64**	−0.05	**−0.15**	**0.50**	**0.69**	**0.71**		
Ex. Ni	**0.35**	**0.21**	**0.26**	0.09	0.06	−0.02	**0.32**	**0.55**	**0.35**	**0.30**	
Ex. Cr	**0.31**	**0.47**	−0.06	**−0.27**	**0.21**	0.06	**−0.23**	0.08	**−0.36**	**−0.22**	**0.37**

P values less than 0.05 are in bold font; *n* = 192; "Ex." is the abbreviation of "Extractable".

Because no heavy metal-containing fertilizers or agricultural or industrial wastes have been applied to these forests, the high concentrations of heavy metals in the soil may be largely attributed to atmospheric deposition [7,8,18]. Values for the annual atmospheric deposition of Zn, Pb, and Cu in the PRD, especially in the urbanized areas, are generally higher than those reported in North America (e.g., the Great Lakes) and Europe (e.g., the North Sea) [7]. A recent study showed that particulate matter with a diameter <2.5 μm ($PM_{2.5}$) in the PRD, especially in Foshan, contained >1.5-times the concentrations of Zn (1360–3260 ng/m^3), Pb (450–1080 ng/m^3), and Cu (190–250 ng/m^3) than the particulate matter in several big cities in both China and other countries (e.g., Beijing in China and Cantabria in Spain) (Zn: 62–600 ng/m^3; Pb: 7.1–320 ng/m^3; Cu: 3.4–70 ng/m^3) [18].

For the non-industrial locations, total soil Zn concentrations (121.9–344.4 mg/kg) were markedly higher than the background value for Guangdong Province (47.3 mg/kg) (Table 4), a finding that agrees with previous reports [20,22]. In another previous study, the rate of atmospheric deposition was found to be five to 111 times higher for Zn (3740–6440 ng/m^3) than for Cu (34–914 ng/m^3) or Pb (147–995 ng/m^3) in this area [8]. These findings suggest that the Zn concentration in soil was strongly influenced by atmospheric Zn deposition in the study area, although it is also possible that the background value of total soil Zn concentration in the PRD is higher than in the other areas of Guangdong Province; this possibility requires further examination.

Total soil Ni and Cr concentrations in the industrial and non-industrial locations were all lower than or similar to background values, suggesting that soil Ni and Cr concentrations have been little influenced by exogenous sources in the studied remnant forests. Although the mean total Mn concentration in all soils was lower than the background value, both total and extractable soil Mn concentrations were significantly higher in the industrial than in the non-industrial locations (Table 4 and Figure 2), suggesting the possibility of Mn contamination of the soils in the industrial locations. This result was consistent with a previous study in which foliar Mn concentrations for 10 woody species were higher in an industrial location than in a non-industrial location in the study area [20]. Close relationships of soil Mn concentrations with soil Pb and Cu concentrations (Figure 4 and Table 6) also suggest that soil has been contaminated with Mn in the study area, given that soil Pb and Cu concentrations appear to have been significantly influenced by exogenous sources.

4.2. Comparisons between Forest Types and among Soil Depths

Total soil Pb, Cu, and Mn concentrations and exchangeable soil Pb and Mn concentrations were all significantly higher in the natural forests than in the plantation forests. This result might be partly explained by the removal of these heavy metals by timber harvest or by the enhanced leaching of heavy metals after timber harvest [10,32,33]. The latter hypothesis is supported by the higher solubility of Cu, Mn, and Pb than of Zn, Ni, and Cr in the study area (Table 5) [22]. It is also possible that the canopy cover is more closed in the remnant natural forests than in the remnant plantation forests, and a closed canopy might reduce the deposition of atmospheric particulates, which would then reduce heavy metal deposition to the soil via throughfall and litterfall [34,35]. The interactions of total soil Pb concentration and extractable soil Mn concentration between location type and forest type (Figure 2) suggest that remnant natural forests at the industrial locations faced the highest potential Pb and Mn contamination.

Assessments of heavy metal pollution of soil usually focus only on surface soils because researchers usually assume that exogenous heavy metals may mainly be retarded in the surface soils [13,14,36]. This assumption was supported by the significant decreases in soil Cu, Zn, Mn, and Pb concentrations with soil depth (Table 3 and Table 4) and by the interactions of total and extractable soil Cu concentrations with location type and soil depth in the present study (Table 3 and Figure 3). Soil Zn and Mn concentrations significantly decreased with soil depth in the extractable forms but not in the total forms, suggesting that exchangeable forms of soil metals were more responsive to external inputs than the total forms [37,38]. However, the accumulation of heavy metals in surface soils may also depend on other factors, such as soil binding capacity, solubility of heavy metals, and plant cycling [11,39,40].

4.3. Implications for Soil Management

Our results from the PRD of China suggest that exogenous sources have resulted in significantly elevated levels of soil Zn in remnant forests at both industrial and non-industrial locations, and in significantly elevated levels of Pb, Cu, and Mn sources in remnant forests at industrial locations. In addition, all of the sampled soils were strongly acidic ($pH_{H2O} < 4.5$), with <20% saturation of base cations [6] and a generally low availability of phosphorus [41]. Phosphate rocks are inexpensive and easy to obtain and may benefit forest growth by immobilizing harmful heavy metals [42] and aluminum [43], alleviating soil acidity [44], and supplying phosphorus and calcium [45]. Therefore, the application of phosphate rock with CaO and P_2O_5 as the main constituents may be cost effective for the remnant forests in the PRD, especially in the industrial areas.

5. CONCLUSIONS

Our results demonstrate that remnant forest soils contain high total Zn concentrations at both industrial and non-industrial locations in the PRD and also contain high total Cu and Pb concentrations at industrial locations, suggesting the potential toxicity effects of Zn, Cu and Pb on vegetation in the remnant forests in the PRD. These patterns agree with previous studies of the levels and compositions of heavy metals in the atmospheric depositions in this region [7,8,18]. Total and/or extractable concentrations of Cu, Pb, and Mn in soils were significantly higher in the natural forests than in the plantation forests, suggesting that the risk of Cu, Pb, and Mn toxicity to vegetation is greater in the natural forests. Based on these results and previous studies, we recommend the application of phosphate rock to the soils in the remnant forests of the PRD.

ACKNOWLEDGMENTS

We are very grateful to two anonymous reviewers for their constructive comments. We also thank Hao Zhang (Fudan University, China) for information on industrial and non-industrial towns in Foshan, China. This research was supported by the Agricultural and Forestry Promotion Project of Nanhai Agro-forestry Extension Centre, Guangdong Province (No. 084101001), the Natural Science Foundation of Guangdong Province (No.8151065005000016), and the National Natural Science Foundation of China (No. 31070409).

AUTHOR CONTRIBUTIONS

Conceived and designed the experiments: Enqing Hou, Huimin Xiang and Dazhi Wen; performed the experiments: Enqing Hou, Huimin Xiang, Jianli Li and Jiong Li; Wrote the paper: Enqing Hou and Dazhi Wen.

REFERENCES

1. Jim, C.Y. Holistic research agenda for sustainable management and conservation of urban woodlands. *Landsc. Urban Plan.* **2011**, *100*, 375–379.

2. Sukopp, H. Human-caused impact on preserved vegetation. *Landsc. Urban Plan.* **2004**, *68*, 347–355.

3. Hu, L.; Li, Z.; Liao, W.B.; Fan, Q. Values of village fengshui forest patches in biodiversity conservation in the Pearl River Delta, China. *Biol. Conserv.* **2011**, *144*, 1553–1559.

4. Gao, H.; Ouyang, Z.; Chen, S.; Koppen, C.S.A. Role of culturally protected forests in biodiversity conservation in Southeast China. *Biodivers. Conserv.* **2013**, *22*, 531–544.

5. Jim, C.Y. Soil recovery from human disturbance in tropical woodlands in Hong Kong. *Catena* **2003**, *52*, 85–103.

6. Hou, E.; Wen, D.; Li, J.; Zuo, W.; Zhang, L.; Kuang, Y.; Li, J. Soil acidity and exchangeable cations in remnant natural and plantation forests in the urbanised Pearl River Delta, China. *Soil Res.* **2012**, *50*, 207–215.

7. Wong, C.S.C.; Li, X.D.; Zhang, G.; Qi, S.H.; Peng, X. Atmospheric deposition of heavy metals in the Pearl River Delta, China. *Atmos. Environ.* **2003**, *37*, 767–776.

8. Lee, C.S.L.; Li, X.; Zhang, L.; Zhang, G.; Peng, X. Biomonitoring of trace metals in the atmosphere using moss (*Hypnum plumaeforme*) in the Nanling Mountains and the Pearl River Delta, Southern China. *Atmos. Environ.* **2005**, *39*, 397–407

9. Rademacher, P. *Atmospheric Heavy Metals and Forest Ecosystems*; Work report of the Institute for World Forestry: Hamburg, Germany, 2013.

10. Berthelsen, B.O.; Steinnes, E. Accumulation patterns of heavy metals in soil profiles as affected by forest clear-cutting.*Geoderma* **1995**, *66*, 1–14.

11. Steinnes, E.; Friedland, A.J. Metal contamination of natural surface soils from long-range atmospheric transport: Existing and missing knowledge. *Environ. Rev.* **2006**, *14*, 169–169.

12. Bergkvist, B.; Folkeson, L.; Berggren, D. Fluxes of Cu, Zn, Pb, Cd, Cr, and Ni in temperate forest ecosystems—A literature review. *Water Air Soil Poll.* **1989**, *47*, 217–286.

13. Pouyat, R.V.; Mcdonnell, M.J. Heavy-metal accumulations in forest soils along an urban-rural gradient in southeastern New York, USA. *Water Air Soil Poll.* **1991**, *57–58*, 797–807.

14. Nygård, T.; Steinnes, E.; Røyset, O. Distribution of 32 elements in organic surface soils: Contributions from atmospheric transport of pollutants and natural sources. *Water Air Soil Poll.* **2012**, *223*, 699–713.

15. Statistics Bureau of Guangdong Province. Main data of the 6th National Census Bulletin of Guangdong Province, 2011. Available online: http://www.gdstats.gov.cn/tjzl/tjgb/201105/t20110511_83329.html (accessed on 9 January 2014). (in Chinese).

16. Weng, Q.H. Land use change analysis in the Zhujiang Delta of China using satellite remote sensing, GIS and stochastic modelling. *J. Environ. Manag.* **2002**, *64*, 273–284.

17. Zhang, H.; Ma, W.C.; Wang, X.R. Rapid urbanization and implications for flood risk management in hinterland of the Pearl River Delta, China: The Foshan study. *Sensors* **2008**, *8*, 2223–2239.

18. Wang, Q.; Tan, J.; Ma, Y.; He, K.; Wei, L.; Yang, F.; Yu, Y.; Wang, J. Characteristics of heavy metals in PM2.5 during winter in Foshan city. *China Environ. Sci.* **2012**, *32*, 1384–1391. (in Chinese).

19. Chen, T.B.; Chen, H.; Tian, Q.Z.; Zheng, Y.M.; Lei, M.; Huang, Z.C.; Wu, X.; Wu, H.T.; Fan, K.K.; Yu, K. Assessment of heavy metal pollution in surface soils of urban parks in Beijing, China. *Chemosphere* **2005**, *60*, 542–551

20. Zhang, L.L.; Wang, H.E.; Li, J.; Kuang, Y.W.; Wen, D.Z. Physiological responses and accumulation of pollutants in woody species under *in situ* polluted condition in Southern China. *J. Plant Res.* **2013**, *126*, 95–103.

21. Wen, D.Z.; Kuang, Y.W.; Liu, S.Z.; Zhang, D.Q.; Lu, Y.D.; Li, J.L. Evidences and implications of vegetation damage from ceramic industrial emission on a rural site in the Pearl River Delta of China. *J. Forest. Res.* **2006**, *17*, 7–12.

22. Sun, F.F.; Wen, D.Z.; Kuang, Y.W.; Li, J.; Zhang, J.G. Concentrations of sulphur and heavy metals in needles and rooting soils of Masson pine (*Pinus massoniana* L.) trees growing along an urban-rural gradient in Guangzhou, China.*Environ. Monit. Assess.* **2009**, *154*, 263–274.

23. China International Urbanization Development Strategy Research Committee. *Report of Urbanization Rate in China*; CIUDSRC: Beijing, China, 2008. (in Chinese)

24. Food and Agriculture Organization of the United Nations—United Nations Educational, Scientific and Cultural Organization. *FAO-UNESCO Soil Map of the World*; (Volume 1); UNESCO: Paris, France, 1974.

25. Liao, Y.H.; Chen, H.Y.; Wang, Z.; Chen, C.G.; Liu, L.W. Study on community of fengshui woods and the value of application in construction of ecological public welfare forest. *J. Subtro. Res. Environ.* **2008**, *3*, 42–48. (in Chinese).

26. Zhuang, X. Floristic composition and conservation of the Fungshui woods in Guangdong's Pearl River Delta and Hong Kong. *Guangdong For. Sci. Technol.* **2012**, *28*, 72–76. (in Chinese).

27. Zhang, H.; Department of Environmental Science and Engineering, Fudan University. Personal Communication. 2014.

28. Soil Classification Working Group (SCWG). *The Canadian System of Soil Classification*; NRC Research Press: Ottawa, ON, Canada, 1998.

29. State Environmental Protection Administration of China (SEPAC). *The Technical Specification for Soil Environmental Monitoring*; HJ/T 166-2004; Environmental Press of China: Beijing, China, 2004. (in Chinese)

30. USEPA. Method 3052: Microwave Assisted Acid Digestion of Siliceous and Organically Based Matrices. US Environmental Protection Agency (USEPA): Washington, DC, USA, 1996.

31. China National Environmental Monitoring Centre. *Background Values of Soil Elements in China*; Environmental Science Press of China: Beijing, China, 1990. (in Chinese)

32. Scott, N.A.; Likens, G.E.; Eaton, J.S.; Siccama, T.G. Trace metal loss following whole-tree harvest of a northeastern deciduous forest, USA. *Biogeochemistry* **2001**, *54*, 197–217.

33. Pulford, I.D.; Watson, C. Phytoremediation of heavy metal-contaminated land by trees—A review. *Environ. Int.* **2003**,*29*, 529–540.

34. Nieminen, T.M.; Derome, J.; Helmisaari, H.S. Interactions between precipitation and Scots pine canopies along a heavy-metal pollution gradient. *Environ. Poll.* **1999**, *106*, 129–137.

35. Brandford, D.; Flower, D.; Moghaddam, M.V. Study of aerosol deposition at a wind exposed forest edge using ^{210}Pb and ^{137}Cs soil inventories. *Water Air Soil Poll.* **2004**, *57*, 107–116.

36. Chen, T.; Wong, J.; Zhou, H.; Wong, M. Assessment of trace metal distribution and contamination in surface soils of Hong Kong. *Environ. Poll.* **1997**, *96*, 61–68.

37. Bermudez, G.M.A.; Moreno, M.; Invernizzi, R.; Plá, R.; Pignata, M.L. Heavy metal pollution in topsoils near a cement plant: The role of organic matter and distance to the source to predict total and HCl-extracted heavy metal concentrations. *Chemosphere* **2010**, *78*, 375–381.

38. Chen, Z.S. Relationship between Heavy Metal Concentrations in Soils of Taiwan and Uptake by Crops. Food & Fertilizer Technology Center: Taipei, Taiwan, 2010.

39. Lee, B.D.; Carter, B.J.; Basta, N.T.; Weaver, B. Factors influencing heavy metal distribution in six Oklahoma benchmark soils. *Soil Sci. Soc. Am. J.* **1997**, *61*, 218–223.

40. Blaser, P.; Zimmermann, S.; Luster, J.; Shotyk, W. Critical examination of trace element enrichments and depletions in soils: As, Cr, Cu, Ni, Pb, and Zn in Swiss forest soils. *Sci. Total Environ.* **2000**, *249*, 257–280.

41. Hou, E.; Chen, C.; Li, J.; Zuo, W.; Wang, Y.; Wang, X.; Wen, D. Soil inorganic phosphorus fractions of remnant native and plantation forests in Foshan region. *J. Trop. Subtrop. Bot.* **2012**, *20*, 546–554.

42. Lin, C.W.; Lian, J.; Fang, H.H. Soil lead immobilization using phosphate rock. *Water Air Soil Poll.* **2005**, *161*, 113–123.

43. Alloush, G.A.; Clark, R.B. Maize response to phosphate rock and arbuscular mycorrhizal fungi in acidic soil. *Commun. Soil Sci. Plant Anal.* **2001**, *32*, 231–254.

44. Rajan, S.S.S.; Watkinson, J.H.; Sinclair, A.G. Phosphate rocks for direct application to soils. *Adv. Agron.* **1996**, *57*, 77–159.

45. Nabahungu, N.; Semoka, J.; Zaongo, C. Limestone, Minjingu phosphate rock and green manure application on improvement of acid soils in Rwanda. In *Advances in Integrated Soil Fertility Management in Sub-Saharan Africa: Challenges and Opportunities*; Springer: Dordrecht, The Netherlands, 2007; pp. 703–712.

Index